Advanced Probability and Statistics: Remarks and Problems

Advanced Probability and Statistics: Remarks and Problems

Harish Parthasarathy
Professor
Electronics & Communication Engineering
Netaji Subhas Institute of Technology (NSIT)
New Delhi, Delhi-110078

CRC Press
Taylor & Francis Group
Boca Raton London New York

CRC Press is an imprint of the
Taylor & Francis Group, an **informa** business

Manakin
PRESS

First published 2023
by CRC Press
4 Park Square, Milton Park, Abingdon, Oxon, OX14 4RN

and by CRC Press
6000 Broken Sound Parkway NW, Suite 300, Boca Raton, FL 33487-2742

CRC Press is an imprint of Informa UK Limited

British Library Cataloguing-in-Publication Data
A catalogue record for this book is available from the British Library

Library of Congress Cataloging-in-Publication Data
A catalog record has been requested

ISBN: 9781032405155 (hbk)
ISBN: 9781032405162 (pbk)
ISBN: 9781003353447 (ebk)

DOI: 10.4324/9781003353447

Typeset in Arial, Calibri, Cambria Math, Century Schoolbook, MT-Extra, Symbol MT, Tahoma, Verdana, Wingdings, Palatino, Monotype Corsiva, Euclid Extra, KozGoPr6N, Minion Pro, Symbol and Times New Roman
by Manakin Press, Delhi

Manakin
PRESS

Preface

The chapters of this book deals with the basic formulation of waveguide cavity resonator equations especially when the cross sections of the guides and resonators have arbitrary shapes. The focus is on expressing the total field energy within such a cavity resonator as a quadratic form in the complex coefficients that determine the modal expansions of the electromagnetic field. Such an expression can then be immediately quantized by replacing the coefficients with creation and annihilation operators.

The reviews of basic statistical signal processing covering linear models, fast algorithms for estimating the parameters in such linear models, applications of group representation theory to image processing problems especially the representations of the permutation groups and induced representation theory applied to image processing problems involving the three dimensional Euclidean motion group. Some attention has been devoted to quantum aspects of stochastic filtering theory. The UKF as an improvement of the EKF in nonlinear filtering theory has been explained.

The Hartree-Fock equations for approximately solving the two electron atomic problem taking spin-orbit magnetic field interactions into account has been discussed. In the limit as the lattice tends to a continuum, the convergence of the stochastic differential equations governing interacting particles on the lattice to a hydrodynamic scaling limit has also been discussed. Statistical performance analysis of the MUSIC and ESPRIT algorithms used for estimating the directions of arrival of multiple plane wave emitting signal sources using an array of sensors has been outlined here. It is based on our understanding of how the singular value decomposition of a matrix gets perturbed when the given matrix is subject to a small random perturbation. Finally, some aspects of supersymmetry and supergravity have been discussed in the light of the fact that supersymmetry is now a mathematically well-defined field of research that has opened up a new avenue to our understanding of how gravity can be unified with the other fundamental forces of nature. This book is based on the lectures delivered by the author to undergraduate and postgraduate students. These courses were on transmission lines and waveguides and statistical signal processing.

Author

Table of Contents

Chapter 1

Remarks and Problems on Transmission Lines and Waveguides

[1] Study about the historical development of the Maxwell equations for electromagnetism starting with the experimental findings and theoretical formulations of Coulomb, Ampere, Oorsted, Faraday, Gauss and finally culminating in Maxwell's introduction of the displacement current to satisfy charge conservation in time varying situations. Study about how Maxwell converted all these findings into laws expressible in the form of partial differential equations based on the basic operations of vector calculus and how by manipulating these equations, he proved that electric and magnetic fields propagate in vacuum as plane waves travelling at the speed of light and thereby how he unified light with electricity and magnetism. Study about how Heinrich Hertz confirmed Maxwell's theory hundred years later using Leyden jar experiments.

Max Planck struggled for over twenty years to finally arrive at his law for the spectrum of black body radiation. The earlier law for this spectrum that was being used was Wien's displacement law according to which the spectral density of black body radiation was proportional to

$$S(\nu) = C\nu^3.exp(-\beta\nu)$$

With $\beta = A/T$ with A a constant. At very low frequencies this law states that the spectral density is proportional to ν^3. The same is true at very high temperatures. At very low temperatures, this law predicts that the spectrum will vanish, ie, there will not be any radiation at all. The high temperature and low frequency limit of Wien's displacement law was in sharp contradiction with experiment. Planck used a little of statistical mechanics but more of curve fitting to modify Wien's displacement law to

$$S(\nu) = \frac{C\nu^3}{exp(A\nu/T) - 1}$$

This law has the same low temperature and high frequency behaviour as Wien's displacement law but at high temperatures or at low frequencies, while Wien's law predicts a $C\nu^3$ dependence of the spectrum of black body radiation, Planck's law gives a behaviour $CT\nu^2$ which is in agreement with the experiments conducted by Rubens and Kurlbaum. Planck by fitting this curve to the experimental curve of Rubens and Kurlbaum arrived at the formula $A = h/k$ where k is Avogadro's number and h is called Planck's constant. Planck later on gave the following derivation of his radiation law: He assumed that radiation energy comes in quanta of $h\nu$, ie, in integer multiples of $h\nu$ via harmonic oscillators. When a harmonic oscillator of frequency ν is excited to the nth energy level, it acquires an energy of $nh\nu$ and by Boltzmann's relation between energy and probability, the probability of such an oscillator getting excited to the nth level is proportional $exp(-nh\nu/kT)$. Hence, Planck concluded that the average energy of an oscillator of frequency ν is given by

$$U(\nu) = \frac{\sum_{n\geq 0} nh\nu.exp(-nh\nu/kT)}{\sum_{n\geq 0} exp(-nh\nu/kT)}$$

$$= \frac{h\nu}{exp(h\nu/kT) - 1}$$

Next, Planck used familiar method of Rayleigh to derive a formula for the total number of oscillators having frequency in the range $[\nu, \nu + d\nu]$ and belonging to the volume spatial V. This number is given by

$$\int_{q\in V, pc\in[h\nu, h(\nu+d\nu]} d^3q d^3p/h^3$$

Where he used Einstein's energy-momentum relation $E = pc$ for photons which have zero mass. Taking into account that the photon has two independent modes of polarization, ie, perpendicular to its direction of propagation, this number evaluates to

$$d\nu\frac{d}{d\nu}V\int_{p\leq h\nu/c} 8\pi p^2 dp/h^3 = d\nu.\frac{d}{d\nu}V 8\pi\nu^3/3c^3 = d\nu.V.8\pi\nu^2/c^3$$

Multiplying this number with the average energy of an oscillator gives us the average energy of black body radiation in the frequency range $[\nu, \nu + d\nu]$:

$$S(\nu)d\nu = \frac{8\pi h\nu^3/c^3}{exp(h\nu/kT) - 1}d\nu$$

This is the famous Planck's law of black body radiation and its advent was the starting point of the whole of modern quantum mechanics and quantum field theory. First Newton came who unified gravitation which causes the apple to fall with Kepler's laws of planetary motion by proposing his celebrated inverse square law, he invented calculus along with Leibniz in order to establish Kepler's laws of motion from his inverse square law of gravitation. More precisely, Robert Hooke who was the curator of the Royal society posed to Newton the inverse problem: What radial force law of attraction should exist between the sun and a planet in order that the planet move around the sun in ellipses satisfying Kepler's laws ? Newton solved this inverse problem by inventing calculus and formulating his second law of motion in terms of differential calculus which he called fluxions. He applied this law to the sun planet system and proved that when the force of attraction between the two is the inverse square law then the planet is guaranteed to revolve around the sun in an ellipse. It is a little unfortunate that Robert Hooke's name does not appear prominently in Newton's magnamopus "Philosophae Naturalis Principia Mathematica' which is the Latin translation of "Mathematical Principles of Natural Philosophy". Today we believe that some portion of the credit for the discovery of the inverse square law of gravitation should go to Robert Hooke.

After Newton, the next major unification in physics came with Maxwell when he created the four laws of electromagnetism based on the findings of Coulomb, Gauss, Ampere, Oorsted and Faraday and using these laws predicted that electricity, magnetism and light are one and the same phenomena which appear distinct phenomena to us primarily because of the frequencies at which these propagate.

[2] The rectangular waveguide:Expressing the transverse component of the electromagnetic field in terms of the longitudinal components.

[a] A rectangular waveguide has dimensions a, b along the x and y axes respectively. Assume that the length of the guide is d. When the fields have the sinusoidal dependence $exp(-j\omega t)$ and dependence upon z as $exp(-\gamma z)$, then $\partial/\partial t$ and $\partial/\partial z$ get replaced respectively by multiplication with $-j\omega$ and $-\gamma$.

Hence, the Maxwell curl equations in the $\omega - x - y - z$ domain are

$$curl\mathbf{E} = -j\omega\mu\mathbf{H}, curl\mathbf{H} = j\omega\epsilon\mathbf{E}$$

which in component form become

$$E_{z,y} + \gamma E_y = -j\omega\mu H_x, - - - - - (1)$$

$$-\gamma E_x - E_{z,x} = -j\omega\mu H_y, - - - - - (2)$$

$$E_{y,x} - E_{x,y} = -j\omega\mu H_z - - - - - (3)$$

and likewise by duality with $\mathbf{E} \to \mathbf{H}, \mathbf{H} \to -\mathbf{E}$, $\epsilon \to \mu, \mu \to \epsilon$. Write down the dual equations:

$$H_{z,y} + \gamma H_y = j\omega\epsilon E_x, - - - - - (4)$$

$$-\gamma H_x - H_{z,x} = j\omega\epsilon E_y, - - - - - (5)$$

$$E_{y,x} - E_{x,y} = j\omega\epsilon E_z - - - - - (6)$$

Solve (1),(2),(4),(5) for $\{E_x, E_y, H_x, H_y\}$ in terms of $\{E_{z,x}, E_{z,y}, H_{z,x}, H_{z,y}\}$ and show that this solution can be expressed as

$$\mathbf{E}_\perp = (-\gamma/h^2)\nabla_\perp E_z - (j\omega\mu/h^2)\nabla_\perp H_z \times \hat{z} - - - (7)$$

$$\mathbf{H}_\perp = (-\gamma/h^2)\nabla_\perp H_z + (j\omega\epsilon/h^2)\nabla_\perp E_z \times \hat{z} - - - (8)$$

where

$$\nabla_\perp = \hat{x}\partial/\partial x + \hat{y}\partial/\partial y,$$

$$\mathbf{E}_\perp = E_x\hat{x} + E_y\hat{y}, \mathbf{H}_\perp = H_x\hat{x} + H_y\hat{y},$$

$$h^2 = \gamma^2 + \omega^2\epsilon\mu$$

[3]

[a] Show that all the procedures and expressions in Step 1 are valid even when ϵ, μ are functions of (ω, x, y) but not of z. Assuming ϵ, μ to be constants, show that by substituting (7) and (8) into (3) and (6) gives us the two dimensional Helmholtz equation for E_z, H_z:

$$(\nabla_\perp^2 + h^2)E_z = 0, (\nabla_\perp^2 + h^2)H_z = 0 - - - (9)$$

[4]

[a] Show using (7) and (8) that the boundary conditions that the tangential components of **E** and the normal components of **H** vanish on its boundary walls are equivalent to the conditions

$$E_z = 0, x = 0, a, y = 0, b, \partial H_z/\partial x = 0, x = 0, a, \partial H_z/\partial y = 0, y = 0, b$$

Hence by applying the separation of variables method to (9) deduce that the general solutions are given in the frequency domain by

$$E_z(\omega, x, y,) = \sum_{m,n\geq 1} c(m,n,\omega)u_{m,n}(x,y)exp(-\gamma_{mn}(\omega)z),$$

$$H_z(\omega, x, y,) = \sum_{m,n\geq 1} d(m,n,\omega)v_{m,n}(x,y)exp(-\gamma_{mn}(\omega)z),$$

where

$$u_{m,n}(x,y) = (2\sqrt{2}/\sqrt{ab}).sin(m\pi x/a).sin(n\pi y/b),$$

$$v_{m,n}(x,y) = (2\sqrt{2}/\sqrt{ab}).cos(m\pi x/a).cos(n\pi y/b),$$

$$\gamma_{m,n}(\omega) = \sqrt{h_{mn}^2 - \omega^2\mu\epsilon},$$

where

$$h_{mn}^2 = ((m\pi/a)^2 + (n\pi/b)^2)^{1/2}$$

where m, n are positive integers. Show that

$$\int_0^a \int_0^b u_{m,n}(x,y)u_{p,q}(x,y)dxdy = \delta_{m,p}\delta_{n,q}$$

and deduce that if $Re(\gamma_{mn}) = \alpha_{mn}$, then

$$\int_0^a \int_0^b \int_0^d |E_z|^2 dxdydz = \sum_{m,n} |c(m,n,\omega)|^2 (1 - exp(-2\alpha_{mn}(\omega)d)/2\alpha_{mn}(\omega)$$

Likewise evaluate

$$\int |H_z|^2 dxdydz, \int |\mathbf{E}_\perp|^2 dxdydz, \int |H_\perp|^2 dxdydz$$

and hence evaluate the time averaged energy density in the electromagnetic field at frequency ω:

$$U = (1/4)\int_0^a \int_0^b \int_0^d (\epsilon|\mathbf{E}(\omega,x,y,z)|^2 + \mu|\mathbf{H}(\omega,x,y,z)|^2)dxdydz$$

Question: Why does the 1/4 factor come rather than the 1/2 factor ?

[5]
[a] Calculate the power dissipated in the waveguide walls assuming that the region outside has a finite conductivity σ?

Solution: The surface current density on the wall $x = 0$ is

$$\mathbf{J}_s(0,y,z) = \hat{x} \times \mathbf{H}(0,y,z) = H_y(0,y,z)\hat{z} - H_z(0,y,z)\hat{y}$$

This is the current per unit length on the wall. It can be attributed to a current density in the infinite region beyond this wall into the boundary having a value of $\mathbf{J}(x,y,z), x < 0$ provided that we take

$$\int_{-\infty}^0 \mathbf{J}(x,y,z)dx = \mathbf{J}_s(0,y,z)$$

However from basic electromagnetic wave propagation theory in conducting media, we know that

$$\mathbf{J}(x,y,z) = \mathbf{J}(0,y,z)exp(\gamma_0 x), \gamma_0 = \sqrt{j\mu\omega(\sigma + j\omega\epsilon)}$$

Thus,

$$\mathbf{J}(0,y,z)/\gamma_0 = \mathbf{J}_s(0,y,z)$$

and hence the average power dissipated inside the region $x < 0, 0 < y < b, 0 < z < d$ is given using Ohm's law by

$$P = \int_{-\infty}^0 \int_0^b \int_0^d (|\mathbf{J}(x,y,z)|^2/2\sigma)dxdydz$$

$$= (\int_0^b \int_0^d |\mathbf{J}_s(0,y,z)|^2/2\sigma|\gamma_0|^2 dydz).(\int_{-\infty}^0 exp(2\alpha_0 x)dx$$

$$= (\int_0^b \int_0^d (|\mathbf{J}_s(0,y,z)|^2/(4\alpha_0\sigma|\gamma_0|^2))dydz)$$

where

$$\alpha_0 = Re(\gamma_0)$$

Repeat this calculation for the other walls.

[6]

[a] Show that if the waveguide has an arbitrary cross section, in an arbitrary orthogonal coordinate system (q_1, q_2) for the $x-y$ plane, equations (7) and (8) can be expressed as

$$E_1 = (-\gamma/h^2G_1)\partial E_z/\partial q_1 - (j\omega\mu/h^2G_2)\partial H_z/\partial q_2$$

$$E_2 = (-\gamma/h^2G_2)\partial E_z/\partial q_2 + (j\omega\mu/h^2G_1)\partial H_z/\partial q_2$$

$$H_1 = (-\gamma/h^2G_1)\partial H_z/\partial q_1 + (j\omega\epsilon/h^2G_2)\partial E_z/\partial q_2$$

$$H_2 = (-\gamma/h^2G_2)\partial H_z/\partial q_2 - (j\omega\epsilon/h^2G_1)\partial E_z/\partial q_2$$

where G_1, G_2 are the Lame's coefficients for orthogonal curvilinear coordinate system (q_1, q_2), ie,

$$G_1 = \sqrt{(\partial x/\partial q_1)^2 + (\partial y/\partial q_1)^2},$$

$$G_2 = \sqrt{(\partial x/\partial q_2)^2 + (\partial y/\partial q_2)^2},$$

and

$$\mathbf{E}_\perp(\omega, q_1, q_2, z) = E_1(\omega, q_1, q_2, z)\hat{q}_1 + E_2(\omega, q_1, q_2, z)\hat{q}_2,$$

$$\mathbf{H}_\perp(\omega, q_1, q_2, z) = H_1(\omega, q_1, q_2, z)\hat{q}_1 + H_2(\omega, q_1, q_2, z)\hat{q}_2,$$

define the curvilinear components of $\mathbf{E}_\perp$ and $\mathbf{H}_\perp$ respectively. Show that combining these equations with the $\hat{z}$ component of the Maxwell curl equations results in the two dimensional Helmholtz equation in the curvilinear system:

$$\frac{1}{G_1G_2}(\frac{\partial}{\partial q_1}\frac{G_2}{G_1}\frac{\partial E_z}{\partial q_1} + \frac{\partial}{\partial q_2}\frac{G_1}{G_2}\frac{\partial E_z}{\partial q_2}) + h^2E_z = 0$$

and the same equation for H_z.

[b] Show that the boundary conditions on the conducting walls in the curvilinear case, assuming that the boundary curve of the waveguide cross section is given by $q_1 = c = constt$ assume the forms

$$E_z = 0, q_1 = c, \frac{\partial H_z}{\partial q_1} = 0, q_1 = c$$

Deduce that in general, the modal eigenvalues h^2 are different in the TE ($H_z = 0$) and the TM ($E_z = 0$) cases. They are the same only in the rectangular waveguide case.

[c] Deduce an expression for the total time averaged power at frequency ω dissipated in the waveguide's conducting walls assuming that the region exterior to the guide has a constant conductivity of σ.

hint: The surface current density is

$$\mathbf{J}_s(\omega, q_2, z) = \hat{q}_1 \times \mathbf{H}_\perp(\omega, c, q_2, z)$$

and hence by the same reasoning as in step 3, we have that the volume current density in the conducting exterior satisfies

$$\nabla^2 \mathbf{J}(\omega, q_1, q_2, z) - \gamma_0(\omega)^2 \mathbf{J}(\omega, q_1, q_2, z) = 0$$

$$\gamma_0(\omega) = \sqrt{j\omega\mu(\sigma + j\omega\epsilon)},$$

$$\int_c^\infty \mathbf{J}(\omega, q_1, q_2, z) G_1(q_1, q_2) dq_1 = \mathbf{J}_s(\omega, q_2, z)$$

An approximate solution to this corresponding to the situation when the field propagates only along the q_1 direction in the conducting region is given by

$$\mathbf{J}(\omega, q_1, q_2, z) = \gamma_0(\omega) \mathbf{J}_s(\omega, q_2, z) exp(-\gamma_0(\omega) \int_c^{q_1} G_1(q_1, q_2) dq_1)$$

Note that this situation corresponds to the fact that the fields propagate from the surface of guide into the depth of the conducting walls normally, ie, along the direction q_1 into it and the fact that in propagating from $q_1 = c$ to q_1 normally, the distance covered is $l = \int_c^{q_1} G_1(q_1, q_2) dq_1$. Then the average power dissipated per in a length d of the guide at frequency ω is given by

$$P_{diss} = (1/2\sigma) \int_c^\infty \int_0^A \int_0^d |\mathbf{J}(\omega, q_1, q_2, z)|^2 G_1(q_1, q_2) G_2(q_1, q_2) dq_1 dq_2 dz$$

where when q_2 varies over $[0, A]$ one full curve on the cross section is covered. Note that q_2 is tangential to the waveguide boundary curve for any cross section.

Specialization to cylindrial guides:

[a] In Step 4, choose $q_1 = \rho = \sqrt{x^2 + y^2}, q_2 = \phi - tan^{-1}(y/x)$Show that (ρ, ϕ) form an orthogonal curvilinear system of coordinates in the xy plane and that

$$G_1 = G_\rho = 1, G_2 = G_\phi = \rho$$

so that equations (7) and (8) assume the forms

$$E_\rho = (-\gamma/h^2) \frac{\partial E_z}{\partial \rho} - (j\omega\mu/h^2\rho) \frac{\partial H_z}{\partial \phi}$$

$$E_\phi = (-\gamma/h^2\rho)\frac{\partial E_z}{\partial \phi} + (j\omega\mu/h^2)\frac{\partial H_z}{\partial \rho}$$

$$H_\rho = (-\gamma/h^2)\frac{\partial H_z}{\partial \rho} + (j\omega\epsilon/h^2\rho)\frac{\partial E_z}{\partial \phi}$$

$$H_\phi = (-\gamma/h^2)\frac{\partial H_z}{\partial \rho} - (j\omega\epsilon/h^2\rho)\frac{\partial E_z}{\partial \phi}$$

[b] Substituting the above into the z component of the Maxwell curl equations then gives us the two dimensional Helmholtz equation for E_z, H_z in the plane polar coordinate system:

$$\frac{1}{\rho}\frac{\partial}{\partial \rho}\rho\frac{\partial E_z}{\partial \rho} + \frac{1}{\rho^2}\frac{\partial^2 E_z}{\partial \phi^2} + h^2 E_z = 0$$

and the same equation for H_z. The boundary conditions are given by

$$E_z = 0, \rho = R, \frac{\partial H_z}{\partial \rho} = 0, \rho = R$$

and these solving them by the method of separation of variables with the application of the appropriate boundary conditions then gives us the general solutions

$$E_z(\omega,\rho,\phi,z) = \sum_{m,n}[J_m(\alpha_m(n)\rho/R)(c_1(\omega,m,n)cos(m\phi)+ c_2(\omega,m,n)sin(m\phi))exp(-\gamma^E_{mn}(\omega)z)]$$

$$H_z(\omega,\rho,\phi,z) = \sum_{m,n}[J_m(\beta_m(n)\rho/R)(d_1(\omega,m,n)cos(m\phi)+ d_2(\omega,m,n)sin(m\phi))exp(-\gamma^H_{mn}(\omega)z)]$$

where $\alpha_m(n), n = 1,2,...$ are the roots of $J_m(x) = 0$ while $\beta_m(n), n = 1,2,...$ are the roots of $J'_m(x) = 0$ and further

$$h = h_{mn} = \alpha_m(n)/R$$

in the TM case ($E_z \neq 0, H_z = 0$) while

$$h = h_{mn} = \beta_m(n)/R$$

in the TE case ($E_z = 0, H_z \neq 0$). Further

$$\gamma^E_{mn}(\omega) = \sqrt{\alpha_m(n)^2/R^2 - \omega^2\mu\epsilon},$$

$$\gamma^H_{mn}(\omega) = \sqrt{\beta_m(n)^2/R^2 - \omega^2\mu\epsilon}$$

are the propagation constants for TM_{mn} and TE_{mn} modes respectively.

Exercises:

[1] Show using the separation of variables applied to the two dimensional Helmholtz equation that $J_m(x)$ actually satisfies the Bessel equation

$$x^2 J_m''(x) + xJ_m'(x) + (x^2 - m^2)J_m(x) = 0$$

[2] Show that if

$$f_{mn}(\rho) = J_m(\alpha_m(n)\rho/R), J_m(\alpha_m(n)) = 0$$

then

$$\rho^2 f_{mn}''(\rho) + \rho f_{mn}'(\rho) + (\alpha_m(n)^2 \rho^2/R^2 - m^2) f_{mn}(\rho) = 0$$

and hence prove the orthogonality relations

$$\int_0^R J_m(\alpha_m(n)\rho/R) J_m(\alpha_m(k)\rho/R)\rho d\rho = 0, n \neq k$$

Hint: Multiply the above differential equation for $f_{mn}(\rho)$ by $f_{mk}(\rho)/\rho$, interchange n and k, subtract the second equation from the first and integrate from $\rho = 0$ to $\rho = R$. Use integration by parts to deduce the identity

$$(\alpha_m(n)^2 - \alpha_m(k)^2) \int_0^R \rho.f_{mn}(\rho).f_{mk}(\rho)d\rho = 0$$

[3] Repeat Exercise [2] with $\alpha_m(n)$ replaced by $\beta_m(n)$ where now $J_m'(\beta_m(n)) = 0$.

[4] Prove the orthogonality relations

$$\int_0^{2\pi} cos(m\phi)cos(n\phi)d\phi = \int_0^{2\pi} sin(m\phi)sin(n\phi)d\phi = 0, m \neq n$$

and

$$\int_0^{2\pi} cos(m\phi)sin(n\phi)d\phi = 0, \forall m, n$$

[5] Using Exercises [2], [3], [4], deduce that the functions

$$J_m(\alpha_m(n)\rho/R)cos(m\phi), J_m(\alpha_m(n)\rho/R)sin(m\phi), m, n = 1, 2, ...$$

are all mutually orthogonal on the disc of radius R, w.r.t to the area measure $\rho.d\rho.d\phi$ and likewise for the functions

$$J_m(\beta_m(n)\rho/R)cos(m\phi), J_m(\beta_m(n)\rho/R)sin(m\phi), m, n = 1, 2, ...$$

[6] Use the result of Exercise [5] to show that when

$$E_z = \sum_{m,n}[J_m(\alpha_m(n)\rho/R)(c_1(\omega,m,n)cos(m\phi)+c_2(\omega,m,n)sin(m\phi))exp(-\gamma^E_{mn}(\omega)z)]$$

then

$$\int_0^R \int_0^{2\pi} |E_z|^2 \rho.d\rho.d\phi =$$

$$\sum_{m,n}\lambda(m,n)(|c_1(\omega,m,n)|^2 + |c_2(\omega,m,n)|^2)exp(-2\alpha^E_{mn}(\omega)z)$$

where

$$\alpha^E_{mn}(\omega) = Re(\gamma^E_{mn}(\omega))$$

and

$$\lambda(m,n) = \int_0^R \int_0^{2\pi} J_m(\alpha_m(n)\rho/R)^2 cos^2(m\phi)\rho.d\rho.d\phi$$

$$= \int_0^R \int_0^{2\pi} J_m(\alpha_m(n)\rho/R)^2 sin^2(m\phi)\rho.d\rho.d\phi$$

$$= \pi \int_0^R J_m(\alpha_m(n)\rho/R)^2 \rho.d\rho$$

Likewise, show that for

$$H_z = \sum_{m,n}[J_m(\beta_m(n)\rho/R)(d_1(\omega,m,n)cos(m\phi)+d_2(\omega,m,n)sin(m\phi))exp(-\gamma^H_{mn}(\omega)z)]$$

we have

$$\int_0^R \int_0^{2\pi} |H_z|^2 \rho.d\rho.d\phi = \sum_{m,n}\mu(m,n)(|d_1(\omega,m,n)|^2+|d_2(\omega,m,n)|^2)exp(-2\alpha^H_{mn}(\omega)z)$$

where

$$\mu(m,n) = \pi \int_0^R J_m(\beta_m(n)\rho/R)^2 \rho.d\rho$$

[7] Prove that

$$\int_0^R J'_m(\alpha_m(n)\rho/R)J'_m(\alpha_m(k)\rho/R)\rho.d\rho = 0, n \neq k$$

hint: Integrate by parts in two different ways and substitute for the second derivatives using Bessel's equation.

[8] Repeat [7] with $\alpha_m(n)$ replaced by $\beta_m(n)$.

[9] Using the results of the previous Exercises and the expressions for $\mathbf{E}_\perp$ and $\mathbf{H}_\perp$ in terms of E_z, H_z to express the time averaged energy in the electric field as

$$U_E = (\epsilon/4)\int_{[0,R]\times[0,2\pi)\times[0,d]} |\mathbf{E}|^2 \rho.d\rho.d\phi.dz$$

$$= (\epsilon/4)\int (|E_z|^2 + |\mathbf{E}_\perp|^2)\rho.d\rho.d\phi.dz$$

$$\sum_{m,n}[p_E(\omega,m,n)(|c_1(\omega,m,n)|^2 + |c_2(\omega,m,n)|^2)$$

$$+q_E(\omega,m,n)(|d_1(\omega,m,n)|^2 + |d_2(\omega,m,n)|^2)]$$

and that in the magnetic field as

$$U_H = (\mu/4)\int_{[0,R]\times[0,2\pi)\times[0,d]} |\mathbf{H}|^2 \rho.d\rho.d\phi.dz$$

$$= (\mu/4)\int (|H_z|^2 + |\mathbf{H}_\perp|^2)\rho.d\rho.d\phi.dz$$

$$\sum_{m,n}[p_H(\omega,m,n)(|d_1(\omega,m,n)|^2 + |d_2(\omega,m,n)|^2)$$

$$+q_H(\omega,m,n)(|c_1(\omega,m,n)|^2 + |c_2(\omega,m,n)|^2)]$$

where $p_E(\omega,m,n), q_E(\omega,m,n), q_E(\omega,m,n), q_H(\omega,m,n)$ depend only on ϵ, μ, R, d as parameters.

Quality factor

[a] The quality factor of a guide is defined as the ratio of the average energy stored per unit length of the guide to the energy dissipated per unit length per cycle.

Exercise

Compute the quality factors for rectangular and cylindrical guides for specified modes TE_{mn} and TM_{mn}.

[b] For a rectangular guide, the wavelength of propagation along the z axis for the TE_{mn} or the TM_{mn} modes when the frequency is more than the cutoff frequency is given by

$$\lambda = 2\pi/\beta_{mn}, \beta_{mn} = j\gamma_{mn} = \sqrt{\omega^2\mu\epsilon - h_{mn}^2}$$

with

$$h_{mn}^2 = (m\pi/a)^2 + (n\pi/b)^2$$

The phase velocity is given by

$$v_{ph} = \nu\lambda = \omega\lambda/2\pi = \omega/\beta_{mn} = \omega/\sqrt{\omega^2\mu\epsilon - h_{mn}^2}$$

This is greater than the speed of light! The phase velocity is therefore not a meaningful measure for the velocity of energy transfer. A more meaningful measure is the group velocity which is based on the following observation. Let a wave field travelling along the z axis be a sum of two harmonic components with a small frequency difference and a small wavelength difference. Thus, it can be expressed as

$$f(t,z) = cos(\omega t - kz) + cos((\omega + \Delta\omega)t - (k + \Delta k)z)$$

This can be in turn expressed using a standard trigonometric identity as

$$f(t,z) = 2cos((\omega + \Delta\omega/2)t - (k + \Delta k/2)z).cos(\Delta\omega t - \Delta kz)$$

The second cosine term represents the slowly varying (in space and time) envelope of the wave while the first cosine term represents the sharp variations of the signal within the envelope. The velocity of energy transfer is measured by that of the envelope and its is given by

$$v_g = \Delta\omega/\Delta k$$

which in the limit becomes

$$v_g = d\omega/dk$$

This called the group velocity. In our case, we find the group velocity is

$$v_g(m,n) = d\omega/d\beta_{mn} = (d\beta_{mn}/d\omega)^{-1} =$$

$$(d\sqrt{\omega^2\mu\epsilon - h_{mn}^2}/d\omega)^{-1} =$$

$$(\mu\epsilon\omega/\beta_{mn})^{-1} = \beta_{mn}/\mu\epsilon\omega$$

$$= \sqrt{1/\mu\epsilon - h_{mn}^2/(\mu\epsilon\omega)^2} < 1/\sqrt{\mu\epsilon}$$

Thus the group velocity is smaller than the velocity of light and is therefore a more meaningful measure of the velocity of energy transfer for the $(m,n)^{th}$ mode.

Energy density in a guide of arbitrary cross section.

[a] Let $u_n(q_1,q_2)$ and $-h_n^2$ be the eigenfunctions and eigenvalues of the Dirichlet problem

$$(\nabla_\perp^2 + h_n^2)u_n(q_1,q_2) = 0, u_n(a,q_2) = 0$$

and let $v_n(q_1,q_2)$ and and $-k_n^2$ be the eigenfunctions and eigenvalues of the Neumann problem

$$(\nabla_\perp^2 + k_n^2)v_n(q_1,q_2) = 0, \frac{\partial v_n(c,q_2)}{\partial q_1} = 0$$

Recall that in the orthogonal coordinate system (q_1,q_2), we have

$$\nabla_\perp^2 = \frac{1}{G_1G_2}(\frac{\partial}{\partial q_1}\frac{G_2}{G_1}\frac{\partial}{\partial q_1} + \frac{\partial}{\partial q_2}\frac{G_1}{G_2}\frac{\partial}{\partial q_2})$$

Exercise: Assuming that the $h_n^{2's}$ are all distinct, show that the $u_n's$ are all orthogonal:

$$\int_D u_n(q_1,q_2)u_m(q_1,q_2)G_1(q_1,q_2)G_2(q_1,q_2)dq_1dq_2 = 0, n \neq m$$

and likewise assuming that the $k_n^{2's}$ are all distinct, show that

$$\int_D v_n(q_1,q_2)v_m(q_1,q_2)G_1(q_1,q_2)G_2(q_1,q_2)dq_1dq_2 = 0, n \neq m$$

where D is the cross-section of the guide parallel to the xy or equivalently q_1q_2 plane.

hint: Use Green's theorem in the form

$$\int_D (u_n\nabla_\perp^2 u_m - u_m\nabla_\perp^2 u_n)G_1G_2dq_1dq_2 =$$

$$\int_\Gamma (u_n\frac{\partial u_m}{\partial q_1} - u_m\frac{\partial u_n}{\partial q_1})G_2dq_2$$

where Γ is the curve bounding the cross section D and is defined by the condition $q_1 = c, z = 0$. Show that the general solution for the longitudinal component of the electric field and the magnetic fields can be expressed as

$$E_z(\omega,q_1,q_2,z) = \sum_n c(\omega,n)u_n(q_1,q_2)exp(-\gamma_n^E(\omega)z),$$

$$H_z(\omega,q_1,q_2,z) = \sum_n d(\omega,n)v_n(q_1,q_2)exp(-\gamma_n^H(\omega)z)$$

where

$$\gamma_n^E(\omega) = \sqrt{h_n^2 - \omega^2\mu\epsilon},$$
$$\gamma_n^H(\omega) = \sqrt{k_n^2 - \omega^2\mu\epsilon}$$

Hence by using formulae of step 4, calculate the transverse curvilinear components of the electromagnetic field. Now prove that the functions $\nabla_\perp u_n(q_1,q_2), n \geq 1$ are also orthogonal, ie,

$$\int (\nabla_\perp u_n, \nabla_\perp u_m)G_1G_2dq_1dq_2 = 0, n \neq m$$

and likewise for $\nabla_\perp v_n$. For this simply apply Green's formula in the form

$$\int_D (u_n\nabla_\perp^2 u_m + (\nabla_\perp u_n, \nabla_\perp u_m))G_1G_2dq_1dq_2$$

$$= \int_\Gamma (u_n\partial u_m/\partial q_1)G_2dq_2 = 0$$

since $u_n = 0$ on Γ. In the case of v_n, use the same formula but with the boundary condition $\partial v_m/\partial q_1 = 0$ on Γ. Also show that

$$\int_D (\nabla_\perp u_n, \nabla_\perp u_n) G_1 G_2 dq_1 dq_2 = h_n^2 \int_D u_n^2 G_1 G_2 dq_1 dq_2 = h_n^2$$

$$\int (\nabla_\perp v_n, \nabla_\perp v_n) G_1 G_2 dq_1 dq_2 = k_n^2 \int_D v_n^2 G_1 G_2 dq_1 dq_2 = k_n^2$$

assuming that the $u_n' s$ and $v_n' s$ are normalized. Finally, prove the orthogonality of $\nabla_\perp u_n \times \hat{z}, n \geq 1$ and of $\nabla_\perp v_n \times \hat{z}, n \geq 1$. Also prove the mutual orthogonality of ∇u_n and $\nabla_\perp v_m \times \hat{z}$ and of $\nabla_\perp u_n \times \hat{z}$ and ∇v_n. For this, you can use the identities

$$((\nabla_\perp u_n \times \hat{z}), (\nabla_\perp u_m \times \hat{z})) = (\nabla_\perp u_n, \nabla_\perp u_m)$$

and likewise for v_n and further with

$$dS = G_1 G_2 dq_1 dq_2$$

observing that

$$\nabla_\perp f = \frac{1}{G_1} f_{,1} \hat{q}_1 + \frac{1}{G_2} f_{,2} \hat{q}_2$$

that

$$\int_D (\nabla_\perp u_n, \nabla_\perp v_m \times \hat{z}) dS$$

$$= \int_D \hat{z}.(\nabla_\perp u_n \times \nabla_\perp v_m) dS$$

$$= \int_D (u_{n,1} v_{m,2} - u_{n,2} v_{m,1}) dq_1 dq_2$$

$$= \int_D ((u_n v_{m,2})_{,1} - (u_n v_{m,1})_{,2}) dq_1 dq_2 = 0$$

again by applying the two dimensional version of the Gauss divergence theorem to the function $(u_n v_{m,2}, -u_n v_{m,1})$ with the boundary condition that u_n vanishes on Γ. In this way using the formulae of step 4, show that the total average energy of the electromagnetic field in the guide can be expressed as

$$U = (\epsilon/4) \int_{D \times [0,d]} (|E_z|^2 + |\mathbf{E}_\perp|^2) dS dz$$

$$+ (\mu/4) \int_{D \times [0,d]} (|H_z|^2 + |\mathbf{H}_\perp|^2) dS dz$$

$$= \sum_n (\lambda(\omega, n) |c(\omega, n)|^2 + \mu(\omega, n) |d(\omega, n)|^2)$$

where $\lambda(\omega, n), \mu(\omega, n)$ are determined completely by h_n^2, k_n^2, d where d is the length of the guide.

Exercise: Calculate the explicit formulae for $\lambda(\omega, n), \mu(\omega, n)$.

[7]
Cavity resonators.

[a] Take a rectangular wave guide of dimensions a, b along the x and y axes respectively and d along the z axis. Cover the bottom $z = 0$ and the top $z = d$ with perfectly conducting plates. We then get a rectangular cavity resonator which is a cuboid with a perfectly conducting boundaries. By applying the waveguide equations of step 1, we get that the $exp(-\gamma z)$ dependence may be replace with any linear combination of $exp(\pm\gamma z)$ and we must choose this linear combination so that H_z vanishes when $z = 0, d$ and E_x, E_y also vanish when $z = 0, d$. It should be noted that the multiplication by $-\gamma$ that replaces $\partial/\partial z$ cannot be done here since we could multiply either by $\pm\gamma$. This means that the cavity resonator case, equations (7) and (8) must be replaced by

$$\mathbf{E}_\perp = -(1/h^2)\frac{\partial}{\partial z}\nabla_\perp E_z - (j\omega\mu/h^2)\nabla_\perp H_z \times \hat{z} - - - (7')$$

$$\mathbf{H}_\perp = (1/h^2)\frac{\partial}{\partial z}\nabla_\perp H_z + (j\omega\epsilon/h^2)\nabla_\perp E_z \times \hat{z} - - - (8')$$

It should be noted that even in the waveguide case, there are two solutions for γ namely $\pm\sqrt{h_{mn}^2 - \omega^2\mu\epsilon}$ and we choose that linear combination of the corresponding exponentials so that with $\mathbf{E}_\perp$ defined by $(7')$, we get that $\mathbf{E}_\perp$ along with H_z vanishes when $z = 0, d$. This conditions are equivalent to H_z and $\partial E_z/\partial z$ vanishing when $z = 0, d$. Note that E_z vanishing when $z = 0, d$ is equivalent to $\frac{\partial}{\partial z}\nabla_\perp E_z = \nabla_\perp \frac{\partial E_z}{\partial z}$ vanishing when $z = 0, d$. Thus we must choose $\gamma = j\pi p/d$ for some integer p and the combination $sin(\pi pz/d) = (exp(\gamma z) - exp(-\gamma z))/2j$ for H_z and $cos(\pi pz/d) = (exp(\gamma z) + exp(-\gamma z))/2$ for E_z. A little speculation will show that this is valid for cavity resonators of arbitrary cross section in the xy plane.

[8]
Exercises

[1] This problem tells us how to analyze the waveguide fields in the presence of a gravitational field which is independent of (t, z) and described in general relativity in terms of an appropriate metric tensor.

Assume that the metric of space-time is diagonal with the coefficients independent of t, z. Thus, the metric has the form

$$d\tau^2 = g_{00}(x, y)dt^2 + g_{11}(x, y)dx^2 + g_{22}(x, y)dy^2 + g_{33}(x, y)dz^2$$

Write down explicitly the components of the Maxwell equations

$$F_{\mu\nu,\sigma} + F_{\nu\sigma,\mu} + F_{\sigma\mu,\nu} = 0, - - -(1)$$

$$(F^{\mu\nu}\sqrt{-g})_{,\nu} = 0 - - - (2)$$

in this background metric assuming the dependence on (t, z) to be of the form

$$F_{\mu\nu}(t,x,y,z) = H_{\mu\nu}(x,y)exp(-j\omega t - \gamma(\omega)z)$$

Identifying equation (1) with the homogeneous Maxwell equations

$$curl\mathbf{E} + j\omega\mathbf{B} = 0, div\mathbf{B} = 0$$

identify the vectors $\mathbf{E}, \mathbf{B}$ in terms of the components of $H_{\mu\nu}$. Now write down the other Maxwell equations in free space (2) in terms of the $H_{\mu\nu}$ and hence in terms of $\mathbf{E}, \mathbf{B}$ and solve for E_x, E_y, B_x, B_y in terms of E_z, B_z and derive the generalized two dimensional Helmholtz equations satisfied by E_z, B_z. In case inhomogeneous permittivity $\epsilon(\omega, x, y)$ and permeability $\mu(\omega, x, y)$ are also to be taken into account in equation (2), first identify (2) with the Maxwell equations

$$div\mathbf{D} = 0, curl\mathbf{H} + j\omega\mathbf{D} = 0$$

and hence determine $\mathbf{D}, \mathbf{H}$ in terms of the components of $F^{\mu\nu}$. Thus, state how the vacuum medium relation

$$F^{\mu\nu} = g^{\mu\alpha}g^{\nu\beta}F_{\alpha\beta}$$

gets modified in the presence of the inhomogeneous medium. Derive therefrom the relationship between $\mathbf{D}, \mathbf{H}$ and $\mathbf{E}, \mathbf{B}$ in the inhomogeneous medium in the presence of this non-flat diagonal metric. Obtain thus the modified generalized Helmholtz equations for E_z, H_z in this inhomogeneous medium in the presence of the above gravitational field. Generalized this theory to the case of orthogonal curvilinear coordinates $\mathbf{q} = (q_1, q_2)$ in the x-y plane, ie, by writing down the metric as

$$d\tau^2 = g_{00}(\mathbf{q})dt^2 + g_{11}(\mathbf{q})dq_1^2 + g_{22}(\mathbf{q})dq_2^2 + g_{12}(\mathbf{q})dq_1dq_2 + g_{33}(\mathbf{q})dz^2$$

[2] Show that the most general solution for the electromagnetic field within a cavity resonator of arbitrary cross section in the xy plane with length d along the z axis is given by

$$E_z(t,q_1,q_2,z) = \sum_{n,p\geq 1} u_n(q_1,q_2).(2/d)^{1/2}.cos(\pi pz/d).Re(c(n,p)exp(-j\omega(n,p)t))$$

$$H_z(t,q_1,q_2,z) = \sum_{n,p\geq 1} v_n(q_1,q_2).(2/d)^{1/2}.sin(\pi pz/d).Re(d(n,p)exp(-j\omega(n,p)t))$$

$$\mathbf{E}_\perp(t,q_1,q_2,z) = \sum_{n,p} h_n^{-2}.(-\pi p/d).\nabla_\perp u_n(q_1,q_2).(2/d)^{1/2}.sin(p\pi z/d).Re(c(n,p).exp(-j\omega^E(n,p)t))$$

$$-\sum_{n,p} k_n^{-2}.(\nabla_\perp v_n)(q_1,q_2)\times\hat{z}).(2/d)^{1/2}.sin(p\pi z/d).Re(j\mu\omega^H(n,p)d(n,p).exp(-j\omega^H(n,p)t))$$

$$\mathbf{H}_\perp(t,q_1,q_2,z)$$
$$=\sum_{n,p} k_n^{-2}.(\pi p/d).\nabla_\perp v_n(q_1,q_2).(2/d)^{1/2}.cos(p\pi z/d).Re(d(n,p).exp(-j\omega^H(n,p)t))$$
$$+\sum_{n,p} h_n^{-2}.(\nabla_\perp u_n)(q_1,q_2)$$
$$\times\hat{z}).(2/d)^{1/2}.cos(p\pi z/d).Re(j\epsilon\omega^E(n,p)c(n,p).exp(-j\omega^E(n,p)t))$$

where the notation of step 6 has been used. Note that the characteristic E-field and H-field frequencies of oscillation are respectively given bu

$$\omega^E(n,p)=(\mu\epsilon)^{-1/2}\sqrt{h_n^2+(p\pi/d)^2},$$

$$\omega^H(n,p)=(\mu\epsilon)^{-1/2}\sqrt{k_n^2+(p\pi/d)^2},$$

To see how these expressions arise, simply use the waveguide formula

$$\omega^2\mu\epsilon+\gamma_n^E(\omega)^2=h_n^2$$

for the transverse magnetic field situations and

$$\omega^2\mu\epsilon+\gamma_n^H(\omega)=k_n^2$$

for transverse electric field situations, combined with the resonator formula (obtained by applying the boundary conditions on the $z=0,d$ surfaces

$$\gamma_n^E(\omega)^2=-(\pi p/d)^2=\gamma_n^H(\omega)^2$$

You must apply the formula

$$\mathbf{E}_\perp=(1/h^2)\frac{\partial}{\partial z}\nabla_\perp E_z$$

$$\mathbf{H}_\perp=(\epsilon/k^2)\frac{\partial}{\partial t}\nabla_\perp E_z\times\hat{z}$$

for transverse magnetic fields, and

$$\mathbf{E}_\perp=(\mu/k^2)\frac{\partial}{\partial t}\nabla_\perp H_z\times\hat{z}$$

$$\mathbf{H}_\perp=(1/k^2)\frac{\partial}{\partial z}\nabla H_z$$

for transverse electric fields and then apply the superposition principle, namely that the total electric field is the superposition over transverse magnetic modes, ie, with $H_z=0$ and the transverse electric modes, ie, with $E_z=0$. These are derived from the formulas of step 1 by replacing $-\gamma$ by $\partial/\partial z$ and $-j\omega$ by $\partial/\partial t$. This is required since the general form of the cavity fields consists not of one mode and one frequency but is rather a superposition over all the modes and cavity frequencies. In other words, for cavity fields we must rewrite the waveguide formulas in the space-time domain (x,y,z,t), rather than in the 2-D space and frequency domain (x,y,ω).

[3] Determine an expression for the total energy per cycle dissipated in a cavity resonator of arbitrary cross section for a $TM_{n,p}$ mode and for a $TE_{n,p}$ mode. By a $TM_{n,p}$ mode, we mean the electric and magnetic fields derived from

$$H_z = 0, E_z = u_n(q_1, q_2)(2/d)^{1/2} cos(\pi pz/d).Re(c(n,p)exp(-j\omega^E(n,p)t)$$

and by a $TE_{n,p}$ mode, we mean the electric and magnetic fields derived from

$$E_z = 0, H_z = v_n(q_1, q_2)(2/d)^{1/2}.sin(\pi pz/d).Re(d(n,p).exp(-j\omega^H(n,p)))$$

Chapter 2

Remarks and Problems on Statistical Signal Processing

[1] Construct the Lattice filter order recursion for an $\mathbb{R}^M$-valued vector stationary stochastic process $\mathbf{X}(t), t \in \mathbb{Z}$ by minimizing

$$\mathbb{E}[\| \mathbf{X}(t) + \sum_{k=1}^{p} \mathbf{A}(k)\mathbf{X}(t-k) \|^2]$$

with respect to the $M \times M$ prediction coefficient matrices $\mathbf{A}(k), k = 1, 2, ..., p$.

hints: Setting the variational derivatives of the above energy w.r.t the $\mathbf{A}(m)'s$ to zero gives us the optimal normal equations in the form of block structured matrix equations

$$\mathbf{R}(m) + \sum_{k=1}^{p} \mathbf{A}_p(k)\mathbf{R}(m-k) = 0, m = 1, 2, ..., p$$

where

$$\mathbf{R}(m) = \mathbb{E}(\mathbf{X}(t)\mathbf{X}(t-m)^T) \in \mathbb{R}^{M \times M}$$

Note that

$$\mathbf{R}(-m) = \mathbf{R}(m)^T$$

Note also that the optimal prediction error covariance is given by

$$\mathbf{R}_e(p) = BbbE(\mathbf{e}_p(t)\mathbf{e}_p(t)^T)$$

where

$$\mathbf{e}_p(t) = \mathbf{X}(t) + \sum_{k=1}^{p} \mathbf{A}(k)\mathbf{X}(t-k)$$

and by virtue of the orthogonality equations or equivalently, the optimal normal equations,

$$\mathbf{R}_e(p) = \mathbf{R}(0) + \sum_{k=1}^{p} \mathbf{A}_p(k)\mathbf{R}(-k)$$

Note that the minimum prediction error energy or p^{th} order is

$$E(p) = Tr(\mathbf{R}_e(t))$$

Now write down the optimal equations in block matrix structured form and apply the block time reversal operator $\mathbf{J}_p$ consisiting of a reverse diagonal having blocks $\mathbf{I}_M$ and with all the other blocks being zero matrices. Note that

$$\mathbf{J}_p\mathbf{R}_p\mathbf{J}_p = \mathbf{S}_p$$

where

$$\mathbf{R}_p = ((\mathbf{R}(k-m)))_{1\leq k,m\leq p}$$

and

$$\mathbf{S}_p = ((\mathbf{R}(m-k)))_{1\leq k,m\leq p}$$

To get at an order recursion, consider a dual normal equation with $\mathbf{A}(k)$ replaced by $\tilde{\mathbf{A}}(k)$ and $\mathbf{R}(k)$ by $\mathbf{S}(k)$.

[2] Consider the RLS lattice algorithm for the multivariate prediction in both order and time. How would you proceed ?

hint: Define the data vector at time N as

$$\mathbf{X}_N = \begin{pmatrix} \mathbf{X}(N)^T \\ \mathbf{X}(N-1)^T \\ .. \\ \mathbf{X}(0)^T \end{pmatrix} \in \mathbb{R}^{(N+1)\times M}$$

and define the data matrix at time N of order p by

$$\mathbf{X}_{N,p} = [z^{-1}\mathbf{X}_N, z^{-2}\mathbf{X}_n, ..., z^{-p}\mathbf{X}_N] \in \mathbb{R}^{(N+1)\times Mp}$$

The optimal matrix predictor at time N of order p is then given by minimizing

$$\| \mathbf{X}_N + \mathbf{X}_{N,p}\mathbf{A}_{N,p} \|^2$$

where $\| \,.\, \|$ denotes the Frobenius norm and

$$\mathbf{A}_{N,p} = \begin{pmatrix} \mathbf{A}_p(1)^T \\ \mathbf{A}_p(2)^T \\ .. \\ \mathbf{A}_p(p)^T \end{pmatrix} \in \mathbb{R}^{Mp\times M}$$

Question: Identify the appropriate Hilbert space for which this problem can be formulated as an orthogonal projection problem.

[3] Calculate the autocorrelation function of the electromagnetic field inside a waveguide of arbitrary cross section assuming that at the feedpoint, namely at the mouth of the guide at the $z = 0$ plane, the correlation of the E_z, H_z fields are known.

[4] How would you develop an EKF for estimating the electromagnetic field in space time over a bounded region from noisy measurements of the same at a finite discrete set of spatial pixels when the driving current density field is a white Gaussian noise field in space-time ?

hint: Write down the wave equation for $\mathbf{A}$ in the form

$$(\nabla^2 - (1/c^2)\partial_t^2)\mathbf{A}(t,\mathbf{r}) = -\mu\mathbf{J}(t,\mathbf{r})$$

Transform it into two first order in time pde's and by spatial pixel discretization, cast it in state variable form. Now, use the fact that the electric and magnetic fields can be expressed in a source free region as

$$\mathbf{E}(t,\mathbf{r}) = -c^2 \int_0^t \nabla \times (\nabla \times \mathbf{A})dt$$

$$\mathbf{B}(t,\mathbf{r}) = \nabla \times \mathbf{A}$$

to arrive at a measurement model for $\partial\mathbf{E}/\partial t, \mathbf{B}$ at a discrete set of spatial pixels. Apply the EKF to this.

[5] Consider the problem of estimating the moments of a vector parameter that modulates a set of potentials for a quantum system. The Hamiltonian is thus of the form

$$H(t,\theta) = H_0 + \sum_{k=1}^{p} \theta(k)V_k(t)$$

The objective is to estimate the moments of the parameters

$$\mu_p(k_1,...,k_p) = <\theta(k_1)...\theta(k_p)>$$

Schrodinger's equation for the wave function is

$$i\psi'(t) = H(t,\theta)\psi(t)$$

and it has a Dyson series solution

$$\psi(t) = U_0(t)\psi(0)+$$

$$\sum_{n=1}^{\infty} \int_{0<t_n<...t_1<t} U_0(t-t_1)V(t_1,\theta)_0U(t_1-t_2)V(t_2,\theta)$$

$$...U_0(t_{n-1}-t_n)V(t_n,\theta)U_0(t_n)\psi(0)dt_1...dt_n$$

where

$$U_0(t) = exp(-itH_0), V(t,\theta) = \sum_{k=1}^{p} \theta(k)V_k(t)$$

The quantum plus classical average of an observable X after time t is given by

$$< X > (t) =<< \psi(t)|X|\psi(t) >>$$

which can be expressed in the form

$$< X > (t) =< X >_0 (t) + \sum_{n\geq,1\leq k_1,...,k_n\leq p} \mu_n(k_1,...,k_n)F(t,k_1,...,k_n) ----(1)$$

where

$$< X >_0 (t) =< U_0(t)\psi(0)|X|U_0(t)\psi(0) >=< \psi(0)|U_0(t)^* X U_0(t)|\psi(0) >$$

Exercise: Derive an explicit formula for $F(t,k_1,...,k_n)$ in terms of $|\psi(0)>$ $, V_k(t), k = 1,...,p$ and $U_0(t)$.

Answer: $F(t,k_1,...,k_n)$ is the coefficient of $\theta(k_1)...\theta(k_n)$ in the sum of terms of the form

$$(\int_{0<t_m<...t_1<t} < \psi(0)|U_0(t-t_1)V(t_1,\theta)_0U(t_1-t_2)V(t_2,\theta)...U_0(t_{m-1}-t_m)$$
$$V(t_m,\theta)U_0(t_m)|dt_1...dt_m)|X$$

$$|(\int_{0<t^r<...t^1<t} < \psi(0)|U_0(t-t_1)V(t_1,\theta)_0U(t_1-t_2)V(t_2,\theta)...U_0(t_{r-1}-t_r)$$
$$V(t_r,\theta)U_0(t_n)|\psi(0) > dt_1...dt_r)$$

where $m + r = n$ and the terms

$$2.Re(\int_{0<t_n<...t_1<t} < \psi(0)|U_0(t-t_1)V(t_1,\theta)_0U(t_1-t_2)V(t_2,\theta)...U_0(t_{n-1}-t_n)$$
$$V(t_r,\theta)U_0(t_n)|XU_0(t)|\psi(0) > dt_1...dt_r)$$

Now derive an RLS lattice algorithm for estimating the parameter moments $\mu_p(k_1,...,k_p)$ recursively in order and in time from the model (1).

[6] Consider the pth order Volterra system

$$y(t) = sum_{k=1}^p \sum_{t_1,...,t_k=0}^{M} h_k(t_1,...,t_k)x(t-t_1)...x(t-t_k) + e(t)$$

By the use of the Kronecker tensor product, cast this equation in the form of a linear model of the type

$$\mathbf{y}_t = \sum_{k=1}^{p} \mathbf{D}_x(t,k,M)\mathbf{h}_{M,k} + \mathbf{e}_t = \mathbf{D}_x(t,M)\mathbf{g}_M + \mathbf{e}_t$$

where $\mathbf{D}_x(t,k,M)$ is a data matrix built out of the input variables $x(s-t_1)...x(s-t_k), s \leq t, t_1,...,t_k = 0,1,...,M$ and

$$\mathbf{D}_x(t,M) = [\mathbf{D}_x(t,1,M),...,\mathbf{D}_x(t,p,M)],$$

$$\mathbf{g}_M = [\mathbf{h}_{M,1}^T, ..., \mathbf{h}_{M,p}^T]^T$$

Derive an RLS lattice algorithm for estimating $\mathbf{g}_M$ recursively in time t and order M for a fixed p (p is the degree of the Volterra system and M is order).

[7] Determine the wave operators for the Hamiltonian pair

$$H_0 = -id/dx, H_1 = -id/dx + V(x)$$

hint:

$$H_1 = exp(-i\int_0^x V(x)dx)(-id/dx).exp(i\int_0^x V(x)dx)$$

and hence

$$exp(itH_1) = exp(-i\int_0^x V(x)dx)exp(td/dx).exp(i\int_0^x V(x)dx)$$

and also make use of Taylor's formula

$$exp(td/dx)f(x) = f(x+t)$$

Then, evaluate the wave operator Ω_+ defined as

$$lim_{t\to\infty}exp(itH_1).exp(-itH_0)f(x)$$

Identify the domain of Ω_+, ie, the set of all functions f for which the above limit exists in $L^2(\mathbb{R})$.

[7] Let H_0, H_1 be two Hamiltonians with spectral measures $E_0(.)$ and $E_1(.)$ respectively. Show that the wave operator acting on a vector $|f>$ is given by

$$\Omega_+|f>= lim_{t\to\infty}exp(itH_1)exp(-itH_0)|f>$$

$$= lim_{t\to\infty}\int exp(it(y-x))dE_1(y).dE_0(x)|f>$$

Now assume that both H_0, H_1 have purely continuous spectra. Then, show that

$$<g|\Omega_+|f>= lim_{t\to\infty}\int_{\mathbb{R}^2} exp(it(y-x)) <g|E_1'(y)E_0'(x)> dxdy$$

$$= lim_{t\to\infty}\int exp(it(y-x))\frac{\partial^2 F_{g,f}(y,x)}{\partial y\partial x}dxdy$$

where

$$F_{g,f}(x,y) =<g|E_1(y)E_0(x)|f>$$

Now suppose that

$$H_{g,f}(z) = \int_{\mathbb{R}} <g|E_1'(z+x)E_0'(x)|f> dx = \int_{\mathbb{R}} \partial_2\partial_1 F_{g,f}(z+x,x)dx$$

exists and is is square integrable in

$$< g|\Omega_+|f >= 0$$

Suppose now that

$$H_1 = H_0 + \epsilon.V$$

where V is a random potential and ϵ is a small perturbation parameter. Show that

$$exp(itH_1) = exp(itH_0)+\epsilon.exp(itH_0)((1-exp(-itad(H_0))/itad(H_0))(V)+O(\epsilon^2)$$
$$= exp(itH_0) + \epsilon.exp(itH_0).g(it.ad(H_0))(V) + O(\epsilon^2)$$

where

$$g(z) = (1 - exp(-z))/z$$

Hence, deduce that

$$\Omega(t) = exp(itH_1).exp(-itH_0) =$$
$$I + \epsilon.exp(itad(H_0))g(it.ad(H_0))(V) + O(\epsilon^2)$$

Hence assuming a given covariance function for the potential V, ie,

$$R_{VV} = \mathbb{E}(V \otimes V)$$

compute

$$\mathbb{E}((\Omega(t) - I) \otimes (\Omega(s) - I))$$

upto $O(\epsilon^2)$ in terms of R_{VV}. Now go one step further in perturbation theory as follows:

$$exp(tH_1) = exp(t(H_0 + \epsilon V)) = exp(tH_0).W(t, \epsilon)$$

say. Then,

$$\partial_t W(t, \epsilon) = \epsilon.exp(-tH_0)V.exp(tH_0)W(t, \epsilon).$$
$$= \epsilon.exp(-t.ad(H_0))(V).W(t, \epsilon)$$

Thus,

$$W(1, \epsilon) = I+\epsilon \int_0^1 exp(-t.ad(H_0))(V)dt$$
$$+\epsilon^2 \int_{0<s<t<1} exp(-t.ad(H_0))(V).exp(-s.ad(H_0))(V)dtds+O(\epsilon^3)$$

Show that

$$\int_0^1 exp(-t.ad(H_0))dt = (1 - exp(-ad(H_0))/ad(H_0),$$

$$\int_{0<s<t<1} exp(-t.ad(H_0)).exp(-s.ad(H_0))dtds$$
$$= \int_0^1 exp(-t.ad(H_0))(1 - exp(-tad(H_0)))dt/ad(H_0)$$

$$= (1 - exp(-ad(H_0))/ad(H_0)^2 - (1 - exp(-2.ad(H_0)))/ad(H_0)$$

$$= g(ad(H_0))/ad(H_0)^2 - 2.g(2.ad(H_0))$$

and hence obtain a formula for

$$\mathbb{E}(\Omega(t) - I) \otimes (\Omega(s) - I))$$

upto $O(\epsilon^3)$.

[8] Let $\mathbf{X}(t), t \in \mathbb{Z}$ be an M-variate zero mean stationary Gaussian stochastic process with autocorrelation

$$\mathbf{R}[k] = \mathbb{E}(\mathbf{X}(t+k)\mathbf{X}(t)^T) \in \mathbb{R}^{M \times M}, k \in \mathbb{Z}$$

Prove that $\mathbf{R}[.]$ is positive semidefinite in the sense that if $\mathbf{z}_1, ..., \mathbf{z}_p \in \mathbb{R}^M$ are arbitrary, then

$$\sum_{k,m=1}^{p} \mathbf{z}_k^T \mathbf{R}[k-m]\mathbf{z}_m \geq 0$$

Prove that this is zero iff

$$\sum_{k=1}^{p} \mathbf{z}_k^T \mathbf{X}[k] = 0$$

ie, the samples of the process are linearly dependent. Now consider the spectral density matrix of the process defined by

$$\mathbf{S}(\omega) = \sum_{k \in \mathbb{Z}} \mathbf{R}[k] exp(-j\omega k) \in \mathbb{C}^{M \times M}, \omega \in \mathbb{R}$$

Define the periodogram spectral density estimate

$$\hat{\mathbf{S}}_N(\omega) = \frac{1}{N}[\sum_{t=1}^{N} \mathbf{X}(t).exp(-j\omega t)].[\sum_{t=1}^{N} \mathbf{X}(t)^T.exp(j\omega t)]$$

Then prove that

$$\mathbb{E}[\hat{\mathbf{S}}_N(\omega)] = \frac{1}{N} \sum_{t,s=1}^{N} \mathbf{R}[t-s].exp(-j\omega(t-s))$$

$$= \sum_{k=-(N-1)}^{N-1} (1 - |k|/N)\mathbf{R}[k] exp(-j\omega k)$$

Prove that if

$$\sum_{k \in \mathbb{Z}} \parallel \mathbf{R}[k] \parallel < \infty,$$

for any matrix norm $\parallel . \parallel$, (note that any two matrix norms on the space of $M \times M$ are equivalent for M finite, ie, they generate the same topology, which means that if the above series converges for any one matrix norm, it then converges for all the matrix norms), then

$$lim_{N \to \infty} \mathbb{E}[\hat{\mathbf{S}}_N(\omega)] = \mathbf{S}(\omega)$$

Now using the formula for stationary zero mean vector values Gaussian processes

$$\mathbb{E}(X_k(t_1)X_l(t_2)X_m(t_3)X_n(t_4))$$

$$= R_{kl}[t_1 - t_2]R_{mn}[t_3 - t_4] + R_{km}[t_1 - t_3]R_{ln}[t_2 - t_4] + R_{kn}[t_1 - t_4]R_{lm}[t_2 - t_3]$$

Obtain a formula for

$$Cov((\hat{\mathbf{S}}_N(\omega_1))_{kl}, (\hat{\mathbf{S}}_N(\omega_2))_{mn}(\omega_2)) =$$

$$\mathbb{E}[\hat{\mathbf{S}}_N(\omega_1)_{kl}(\hat{\mathbf{S}}_N(\omega_2))_{mn}] -$$

$$\mathbb{E}[\mathbf{S}_N(\omega_1)_{kl}].\mathbb{E}[(\hat{\mathbf{S}}_N(\omega_2))_{mn}]$$

and show that this does not converge to zero as $N \to \infty$. What does it converge to and what can you infer about this result ?

[9] This problem outlines the steps for proving the singular value decomposition of an $M \times N$ complex matrix $\mathbf{A}$.

step 1: Show that $\mathbf{P} = \mathbf{A}^*\mathbf{A}$ is an $N \times N$ positive semidefinite matrix and hence it can be diagonalized by the spectral theorem as

$$\mathbf{P} = \mathbf{UDU}^*$$

where $\mathbf{U}$ is an $N \times N$ unitary matrix and

$$\mathbf{D} = diag[\sigma_1^2, ..., \sigma_r^2, 0, ..., 0], \sigma_1, ..., \sigma_r > 0$$

is a diagonal $N \times N$ matrix with exactly r positive entries and all the other as zero entries where $r = rank(\mathbf{A}) = rank(\mathbf{P}) = rank(\mathbf{Q})$ where $\mathbf{Q} = \mathbf{AA}^*$.

step 2: Let

$$\mathbf{U} = [\mathbf{u}_1, ..., \mathbf{u}_N]$$

Then show that

$$\mathbf{v}_k = \mathbf{Au}_k/\sigma_k, k = 1, 2, ..., r$$

form an orthonormal set of r vectors in $\mathbb{C}^M$. Note that $\mathbf{u}_1, ..., \mathbf{u}_N$ are orthonormal in $\mathbb{C}^N$. Now extend the orthonormal set $\{\mathbf{v}_1, ..., \mathbf{v}_r\}$ to an orthonormal set $\{\mathbf{v}_1, ..., \mathbf{v}_M\}$ in $\mathbb{C}^M$. Show that $\{\mathbf{v}_1, .., \mathbf{v}_r\}$ forms an orthonormal basis for $\mathcal{R}(\mathbf{A})$ while $\{\mathbf{v}_{r+1}, .., \mathbf{v}_M\}$ forms an orthonormal basis for $\mathcal{R}(\mathbf{A})^\perp = \mathcal{N}(\mathbf{A}^*)$. In fact, show that

$$\mathbf{A}^*\mathbf{v}_k = \sigma_k\mathbf{u}_k, k = 1, 2, ..., r,$$

$$\mathbf{A}^*\mathbf{v}_k = 0, k = r + 1, ..., M$$

step 3:Now define the $M \times M$ unitary matrix

$$\mathbf{V} = [\mathbf{v}_1, ..., \mathbf{v}_M]$$

Then show that the discussion in step 2 can be expressed as

$$\mathbf{A}^*\mathbf{V} = \mathbf{U}\Sigma^T$$

where Σ^T is an $N \times M$ matrix of the form

$$\Sigma^T = \begin{pmatrix} \Sigma_r & \mathbf{0} \\ \mathbf{0} & \mathbf{0} \end{pmatrix}$$

where

$$\Sigma_r = diag[\sigma_1, .., \sigma_r]$$

Conclude that

$$\mathbf{A}^* = \mathbf{U}\Sigma.\mathbf{V}^*$$

and hence

$$\mathbf{A} = \mathbf{V}\Sigma.\mathbf{U}^*$$

[10] Using the singular value decomposition for rectangular matrices described in problem [9], obtain the general solution to the least squares problem of computing a vector θ such that $\| \mathbf{x} - \mathbf{A}\theta \|^2$ is a minimum. Also determine that least squares solution θ having minimum norm and prove that it is given by

$$\theta = pinv(\mathbf{A})\mathbf{x}$$

where

$$pinv(\mathbf{A}) = \mathbf{U}\Gamma\mathbf{V}^*$$

where Γ is the $N \times M$ matrix defined by

$$\Gamma = \begin{pmatrix} \Sigma_r^{-1} & \mathbf{0} \\ \mathbf{0} & \mathbf{0} \end{pmatrix}$$

[11] In this problem, we outline a step-wise procedure for statistical image processing on curved surface on which a Lie group of transformations acts. The ideas are based on group representation theory.

step 1: Let $\mathcal{M}$ be a set on which a Lie group G of transformations acts transitively. By a group action, we mean a map

$$\tau : G \times \mathcal{M} \to \mathcal{M}$$

where by writing

$$\tau(g, x) = g.x, g \in G, x \in \mathcal{M}$$

we have

$$(g_2g_1).x = g_2.(g_1.x), g_1, g_2 \in G, x \in \mathcal{M}$$

By a transitive group action, we mean that given any two $x, y \in \mathcal{M}$, we can find a $g \in G$ so that $y = g.x$.

step 2: Let dg be a left invariant Haar measure for the group G, ie, $d(h.g) = dg, h \in G$ or more precisely,

$$\int_G f(h.g)dg = \int_G f(g)dg, h \in G$$

Choose and fix an $x_0 \in \mathcal{M}$ and define a map

$$\tau : G \to \mathcal{M}$$

by

$$\tau(g) = gx_0, g \in G$$

Note that by transitivity of the group action, $\tau(g)$ covers the whole of $\mathcal{M}$ as g varies over G, ie, τ is surjective:$\tau(G) = \mathcal{M}$. Prove that if we denote the left invariant Haar measure on G by $d\mu(g)$, then $d\nu(x) = d\mu o \tau^{-1}(x)$ is an invariant measure on $\mathcal{M}$. Apply this argument to the special case of $SO(3)$ acting on the unit sphere S^2 to prove that that the measure induced on S^2 by the Haar measure on $SO(3)$ is the area measure $sin(\theta)d\theta.d\phi$.

step 3: Construction of the irreducible representations of $SO(3)$ using spherical harmonics. The rotation group $SO(3)$ acts on the unit sphere

$$S^2 = \{\mathbf{x} \in \mathbb{R}^3 : \| \mathbf{x} \| = 1\}$$

This action induces an action on the Hilbert space $L^2(S^2) = \{f : S^2 \to \mathbb{C} : \int_{S^2} |f(\mathbf{x})|^2 dS(\mathbf{x}) < \infty\}$ in a natural way, ie for $g \in SO(3)$, $U(g) : L^2(S^2) \to L^2(S^2)$ is defined by

$$U(g)f(\mathbf{x}) = f(g^{-1}\mathbf{x}), \mathbf{x} \in S^2$$

Prove that U is a representation of $SO(3)$, ie,

$$U(\mathbf{I}_3) = \mathbf{I}_{L^2(S^2)}, U(g_1)U(g_2) = U(g_1g_2), g_1, g_2 \in SO(3)$$

To do image processing on the sphere, we must decompose the representation U into irreducibles. The first step is to note that the Lie algebra of $SO(3)$ is the set of all real 3×3 skew symmetric matrices. This Lie algebra has a standard basis $\{iL_1, iL_2, iL_3\}$ satisfying the commutation relations

$$[L_1, L_2] = iL_3, [L_2, L_3] = iL_1, [L_3, L_1] = iL_2$$

The differential dU of the representation U of $SO(3)$ is a representation of the Lie algebra $\mathfrak{so}(3)$ of $SO(3)$ in $L^2(S^2)$. Its action is given by

$$dU(\mathbf{A})f(\mathbf{x}) = \frac{d}{dt} f(exp(-t\mathbf{A})\mathbf{x})|_{t=0}, \mathbf{A} \in \mathfrak{so}(3)$$

Equivalently,

$$U(exp(t\mathbf{A})) = exp(t.dU(\mathbf{A})), t \in \mathbb{R}, \mathbf{A} \in \mathfrak{so}(3)$$

Note that

$$dU(\mathfrak{so}(3)) = \{dU(\mathbf{A}) : \mathbf{A} \in \mathfrak{so}(3)\}$$

is a three dimensional Lie algebra defined by the commutation relations

$$[dU(\mathbf{A}), dU(\mathbf{B})] = dU([\mathbf{A}, \mathbf{B}]), \mathbf{A}, \mathbf{B} \in \mathfrak{so}(3)$$

We write

$$\tilde{L}_k = dU(\mathbf{L}_k), k = 1, 2, 3$$

and hence

$$[\tilde{L}_1, \tilde{L}_2] = i\tilde{L}_3, [\tilde{L}_2, \tilde{L}_3] = -\tilde{L}_1,$$
$$[\tilde{L}_3, \tilde{L}_1] = i\tilde{L}_2$$

$\tilde{L}_k, k = 1, 2, 3$ are known as the angular momentum operators in quantum mechanics. Decomposing U into irreducibles is therefore equivalent to decomposing the Lie algebra generated by the differential operators $\tilde{L}_k, k = 1, 2, 3$ acting in $L^2(S^2)$ into irreducibles.

Remark: Note that

$$i\tilde{L}_k f(\mathbf{x}) = \frac{d}{dt} f(exp(-itL_k)\mathbf{x})|_{t=0}, k = 1, 2, 3$$

Noting that $exp(itL_1) \subset SO(3)$ is a rotation around the x axis by the angle t and likewise $exp(itL_2), exp(itL_3)$ are respectively rotations around the y and z axis by an angle t, deduce that $\tilde{L}_1 = -i(y\partial/\partial z - \partial/\partial y)$, $\tilde{L}_2 = -i(z\partial/\partial x - x\partial/\partial z)$, $\tilde{L}_3 = -i(x\partial/\partial y - y\partial/\partial x)$ all restricted to $L^2(S^2)$.

Exercise (a): Calculate the explicit forms of the differential operators $\tilde{L}_k, k = 1, 2, 3$ in terms of the spherical polar coordinates on the unit sphere

S^2 $\theta, \phi, \partial/\partial\theta, \partial/\partial\phi$.

(b) Define the second order differential operator $\tilde{L}^2 = \sum_{k=1}^{3} \tilde{L}_k^2$. Prove using the commutation relations for the $\tilde{L}_k's$, that $\tilde{L}^2$ is a second order differential operator that commutes with $\tilde{L}_k, k = 1, 2, 3$. It is known as the Casimir operator for the Lie algebra $dU(\mathfrak{so}(3))$.

(c) Prove that $\tilde{L}^2$ is the negative of the angular part of the Laplacian in three dimensions and is therefore a self-adjoint operator in the Hilbert space $L^2(S^2)$.

(d) From (b) and (c) and the spectral theorem for self-adjoint operators in a Hilbert space, deduce that the eigenspaces of $\tilde{L}^2$ are left invariant under $\tilde{L}_k, k = 1, 2, 3$. In particular, show that $\{\tilde{L}^2, \tilde{L}_3\}$ are jointly diagonable and show by separation of the θ, ϕ variables that these joint eigenfunctions are the spherical harmonics $Y_{lm}(\mathbf{x}), \mathbf{x} = (\theta, \phi) \in S^2$ satisfying the eigen-equations

$$\tilde{L}^2 = Y_{lm} = l(l+1)Y_{lm}, \tilde{L}_3 Y_{lm} = mY_{lm}, m = -l, -l+1, ..., l-1, l, l = 0, 1, 2, ...$$

By the spectral theorem, $\{Y_{lm} : |m| \leq l, l = 0, 1, ...\}$ form a complete orthonormal basis for $L^2(S^2)$.

(e) Define the ladder operators

$$\tilde{L}_+ = \tilde{L}_1 + i\tilde{L}_2, \tilde{L}_- = \tilde{L}_1 - i\tilde{L}_2$$

Prove that

$$[\tilde{L}_+, \tilde{L}_3] = -i\tilde{L}_2 - i\tilde{L}_1 = -\tilde{L}_+,$$
$$[\tilde{L}_-, \tilde{L}_3] = -i\tilde{L}_2 + \tilde{L}_1 = \tilde{L}_-$$

or equivalently,

$$\tilde{L}_+(\tilde{L}_3 + 1) = \tilde{L}_3\tilde{L}_+, \tilde{L}_-(\tilde{L}_3 - 1) = \tilde{L}_3\tilde{L}_-$$

Hence, verify that

$$(m + 1)\tilde{L}_+Y_{lm} = \tilde{L}_3\tilde{L}_+Y_{lm},$$
$$(m - 1)\tilde{L}_-Y_{lm} = \tilde{L}_3\tilde{L}_-Y_{lm}$$

Also $\tilde{L}^2$ commutes with $\tilde{L}_+, \tilde{L}_-$ and hence

$$\tilde{L}^2\tilde{L}_+Y_{lm} = l(l + 1)\tilde{L}_+Y_{lm},$$

$$\tilde{L}^2\tilde{L}_-Y_{lm} = l(l + 1)\tilde{L}_-Y_{lm}$$

Verify that upto a proportionality constant, there is just one function, ie, Y_{lm} which is simulteneously an eigenfunction of $\tilde{L}^2$ with eigenvalue $l(l + 1)$ and an eigenfunction of $\tilde{L}_3$ with eigenvalue m and hence conclude that

$$\tilde{L}_+Y_{lm} = c(l, m)Y_{l,m+1}, \tilde{L} - Y_{lm} = d(l, m)Y_{l,m-1}$$

for some complex constants $c(l, m), d(l, m)$. Note that Y_{lm} is to be interpreted as zero if $m < -l$ or $m > l$. Assuming that the $Y_{lm}'s$ are normalized in $L^2(S^2)$, verify their orthogonality from basic properties of eigenfunctions of self-adjoint operators in a Hilbert :

$$< Y_{lm}, Y_{l'm'} >= \int_{S^2} \bar{Y}_{lm}(\theta, \phi)Y_{l'm'}(\theta, \phi)sin(\theta)d\theta, d\phi = \delta_{ll'}\delta_{mm'}$$

Conclude that $\{Y_{lm} : |m| \leq l\}$ in an orthonormal basis for a subspace V_l of $L^2(S^2)$ that is invariant under $\tilde{L}_k, k = 1, 2, 3$ and hence under $U(SO(3))$.

(e) Prove that V_l has no non-trivial subspaces that are invariant under $\tilde{L}_k, k = 1, 2, 3$, ie, the restriction of $U(SO(3))$ to V_l is an irreducible representation of $SO(3)$. Denote this representation by π_l.
hint: Use the properties of $\tilde{L}_3$ and $\tilde{L}_+, \tilde{L}_-$ acting on Y_{lm}.

(f) Prove that $\pi_l, l = 0, 1, 2, ...$ exhaust all the inequivalent irreducible representations of $SO(3)$.

hint: For this, you must make use of the Peter-Weyl theorem which states that if G is a compact group, then $\pi_l, l = 1, 2, ...$ are all the inequivalent irreducible unitary representations of G iff any $f \in L^2(G)$ can be expanded as

$$f(g) = \sum_l d(l)Tr(f(l)\pi_l(g)), f(l) = \int_G f(g)\pi_l(g)^* dg$$

iff

$$\delta_e = \sum_l d(l)\chi_l(g), \chi_l(g) = Tr(\pi_l(g))$$

iff for any class function $f(g)$ on G (By a class function, we mean any function that satisfies $f(hgh^{-1}) = f(g)\forall g, h \in G$), the relation

$$\int_G barf(g)\chi_l(g)dg = 0, \forall l$$

implies $f = 0$. χ_l is called the character of the representation π_l. Prove that if π_l is the restriction of $U(SO(3))$ to V_l, then

$$\chi_l(R_z(\psi)) = \sum_{m=-l}^{l} exp(-im\psi) = exp(il\psi)(exp(-i(2l+1)\psi) - 1)/exp(-i\psi) - 1)$$

$$= sin((l + 1/2)\psi)/sin(\psi/2), l = 0, 1, 2, ...$$

using the fact that

$$\tilde{L}_3 Y_{lm} = mY_{lm}$$

and hence

$$R_z(\psi)Y_{lm} = exp(-i\psi\tilde{L}_3)Y_{lm} = exp(-im\psi)Y_{lm}$$

Note that χ_l is a class function since

$$Tr(\pi_l(hgh^{-1})) = Tr(\pi_l(h)\pi_l(g)\pi_l(h)^{-1}) = Tr(\pi_l(g))$$

Now let f be a class function on $SO(3)$. Calculate the Haar measure on $SO(3)$ in the form

$$d\mu(g) = F(\theta, \phi, \psi)d\theta.d\phi.d\psi, g = R.R_z(\psi).R^{-1}$$

where

$$R = R_y(\theta)R_z(\phi)$$

and hence evaluate for a class function f,

$$\int_{SO(3)} f(g)\chi_l(g)dg = \int_0^{2\pi} f(R_z(\psi))\chi_l(R_z(\psi))F_0(\psi)d\psi$$

where

$$F_0(\psi) = \int_0^{\pi}\int_0^{2\pi} F(\theta, \phi, \psi)d\theta.d\phi$$

Deduce that if this vanishes for all $l = 0, 1, 2, ...$, then $f = 0$ proving the completeness of the irreducible representations π_l constructed as restrictions of $U(SO(3))$ to $V_l = span(Y_{lm} : |m| \leq l\}$.

(g) Construction of the left invariant Haar measure on a locally compact Lie group G. Let $\omega_1, ..., \omega_n$ be a basis of left invariant one forms on G. Then in component form, we have

$$\omega_k(X)(g) = \omega_k^m(g)X_m(g)$$

with summation over the index m where $X_m(g)$ are the components of a left invariant vector field X at g. We have

$$D(g) = det((\omega_k^m(g))) = (\omega_1 \wedge ... \wedge \omega_n)(g)$$

Now let L_g denote left translation on G, ie, $L_g h = gh, g, h \in G$. Then by the definition of left invariant vector fields and left invariant one forms, we have

$$dL_g X(e) = X(g), (dL_{g^{-1}})^* \omega_k(e) = \omega_k(g)$$

Note that

$$(dL_{g^{-1}})^* = (dL_g)^{*-1}$$

Thus,

$$\omega_k(g) = \omega_k(e)o(dL_g)^{-1}$$

and hence

$$\omega_k(g)(X(g)) = \omega_k(X)(g) = (\omega_k(e)o(dL_g)^{-1})(dL_g X(e)) = \omega_k(e)(X(e))$$
$$= \omega_k(X)(e)$$

is independent of $g \in G$, ie, it is a constant. Now, let $g = (g_1, ..., g_n)$ denote coordinates for $g \in G$. Then,

$$\int_G f(h^{-1}g)D(g)dg_1...dg_n = \int f(g)D(hg)(detdL_h)dg_1...dg_n$$

where L_h the left translation by h represented in this coordinate system and hence dL_h becomes the $n \times n$ Jacobian matrix of L_h. We have on the other hand,

$$D(hg) = (\omega_1 \wedge ... \wedge \omega_n)(hg) =$$
$$(dL_h)^{*-1}(\omega_1 \wedge ... \wedge \omega_n)(g)$$
$$= ((dL_h)^{*-1}\omega_1 \wedge ... \wedge (dL_h)^{*-1}\omega_n)(g)$$
$$= ((dL_h)^{-1*}\omega_1 \wedge ... \wedge (dL_h)^{-1*}\omega_n)(g)$$
$$= \omega_1(g)o(dL_h)^{-1} \wedge ... \wedge \omega_n(g)o(dL_h)^{-1}$$
$$= det(dL_h)^{-1}(\omega_1 \wedge ... \wedge \omega_n)(g)$$

$$= det(dL_h)^{-1}D(g)$$

and we get the required invariance result:

$$\int_G f(h^{-1}g)D(g)dg_1...dg_n = \int_G f(g)D(g)dg_1...dg_n, \forall h \in G$$

Further, we have that if $X^1, ..., X^n$ is any basis for the space of left invariant vector fields on G (ie a basis for the Lie algebra of G) and if $\omega_1, ..., \omega_n$ is it dual basis, then $\omega_1, ..., \omega_n$ is a basis for the space of left invariant one forms on G. Left invariance of the latter follows from the fact that

$$\omega_k(g)(X^m(g)) = \delta_k^m$$

by hypothesis and

$$X^k(g) = dL_g X^k(e)$$

so that

$$\delta_k^m = \omega_k(g)(dL_g X^m(e)) = ((dL_g)^* \omega_k(g))(X^m(e)) = \omega_k(e)(X^m(e))$$

and hence since $X^k(e), k = 1, 2, ..., n$ form a basis for the tangent space to G at e (ie, of the Lie algebra of G), it follows that

$$(dL_g)^* \omega_k(g) = \omega_k(e)$$

or equivalently,

$$\omega_k(g) = (dL_g)^{1*}\omega_k(e)$$

proving left invariance of the one forms $\omega_k's$. Now writing the equation

$$\omega_k(g)(X^m(g)) = \delta_k^m$$

in terms of components, we get

$$\omega_k^r(g) X_r^m(g) = \delta_k^m$$

where summation is over $r = 1, 2, ..., n$ and hence taking determinants on both sides, we get

$$D(g).det((X_r^m(g))) = 1$$

or equivalently,

$$D(g) = \frac{1}{det((X_r^m(g)))}$$

In other words, the left invariant Haar density $D(g)$ for the group G in any given coordinate system for G is just the reciprocal of the wedge product of a basis of left invariant vector fields on G, or equivalently of the determinant of the matrix formed by taking the components of n left invariant vector fields. This in fact gives a nice algorithm for computing the Haar measure on a matrix lie group.

The algorithm:

step 1: Let G be a matrix Lie group of dimension n. Choose a basis $\{X_1, ..., X_n\}$ for the Lie algebra of G. Represent any $g \in G$ as

$$g = exp(t_1 X_1)...exp(t_n X_n), t_1, ..., t_n \in \mathbb{R}$$

For each $k = 1, 2, ..., n$, and $t \in \mathbb{R}$ compute

$$g.exp(tX_k) = exp(t_1 X_1)...exp(t_n X_n).exp(tX_k)$$

and express it as

$$g.exp(tX_k) = exp(a_{k1}(t, t_1, ..., t_n)X_1)...exp(a_{kn}(t, t_1, ..., t_n)X_n), k = 1, 2, ..., n$$

For a function f on G, we write

$$\tilde{f}(t_1, ..., t_n) = f(g) = f(exp(t_1 X_1)...ep(t_n X_n))$$

and then observe that

$$f(g.exp(tX_k)) = \tilde{f}(a_{k1}(t, t_1, ..., t_n), a_{k2}(t, t_1, ..., t_n), .., a_{kn}(t, t_1, ..., t_n))$$

and hence if we use the notation $\tilde{X}_k$ for the differential operator associated with the vector field X_k, we get

$$(\tilde{X}_k f)g) = \frac{d}{dt} f(g.exp(tX_k))|_{t=0} =$$

$$\sum_{m=1}^{n} \frac{\partial a_{km}(0, t_1, ..., t_n)}{\partial t} \frac{\partial \tilde{f}(t_1, ..., t_n)}{\partial t_m}$$

This means that in the above set of coordinates $(t_1, ..., t_n)$ on G, the left invariant vector field X_k is represented by the first order differential operator

$$\tilde{X}_k = \sum_{m=1}^{n} \frac{\partial a_{km}(0, t_1, ..., t_n)}{\partial t} \frac{\partial}{\partial t_m}$$

and therefore the left invariant Haar density $D(t_1, ..., t_n)$ on G in the coordinate system $(t_1, ..., t_n)$ is given by

$$D(t_1, ..., t_n)^{-1} = det((a_{km}(0, t_1, ..., t_n)))_{1 \le k, m \le n}$$

Note that the corresponding left invariant Haar integral $\int_G f(g)dg$ is expressed in the above coordinate system as

$$\int f(exp(t_1 X_1)...exp(t_n X_n)) D(t_1, ..., t_n) dt_1...dt_n$$

$$= \int \tilde{f}(t_1, .., t_n) D(t_1, ..., t_n) dt_1...dt_n$$

Example: Here we compute the Haar measure on $SO(3)$ in terms of the Euler angles. Any rotation R (ie, $R \in SO(3)$) can be represented as

$$R = R_z(\phi)R_x(\theta)R_z(\psi)$$

ϕ, θ, ψ are known as the Euler angles. Note that

$$R_z(\phi) = exp(-iL_3\phi), R_x(\theta) = exp(-iL_1\theta)$$

We write $X_k = -iL_k, k = 1, 2, 3$. Then,

$$R_z(\phi) = exp(\phi X_3), R_x(\theta) = exp(\theta X_1)$$

and

$$[X_1, X_2] = -[L_1, L_2] = -iL_3 = X_3, [X_2, X_3] = X_1, [X_3, X_1] = X_2$$

Note that X_1, X_2, X_3 are real skew-symmetric matrices. Then, write

$$\tilde{f}(\phi, \theta, \psi) = f(R) = f(exp(\phi.X_3).exp(\theta.X_1).exp(\psi.X_3))$$

We find that the left invariant vector fields $\tilde{X}_k, k = 1, 2, 3$ associated with the $SO(3)$ Lie algebra basis elements $X_k, k = 1, 2, 3$ respectively are given by the following computations:

$$\tilde{X}_3\tilde{f}(\phi, \theta, \psi) = \frac{d}{dt}f(exp(\phi.X_3).exp(\theta.X_1).exp(\psi.X_3).exp(tX_3))|_{t=0}$$

$$= f(exp(\phi.X_3).exp(\theta.X_1).exp(\psi.X_3)X_3) = \frac{\partial}{\partial\psi}\tilde{f}(\phi, \theta, \psi)$$

$$\frac{\partial}{\partial\theta}\tilde{f}(\phi, \theta, \psi) =$$

$$f(R_z(\phi)R_x(\theta)X_1R_z(\psi)) =$$

$$f(R_z(\phi)R_x(\theta)R_z(\psi)R_z(-\psi)X_1R_z(\psi))$$

$$= f(R.exp(-\psi.ad(X_3))(X_1))$$

Now,

$$exp(-\psi.ad(X_3))(X_1) = X_1 - \psi.[X_3, X_1] + (\psi^2/2)[X_3, [X_3, X_1]) + ...$$

$$= X_1 - \psi.X_2 - (\psi^2/2)X_1 + ... =$$

$$X_1.cos(\psi) - X_2.sin(\psi)$$

and hence,

$$\frac{\partial\tilde{f}(\phi, \theta, \psi)}{\partial\theta} =$$

$$f(R.(cos(\psi)X_1 - sin(\psi).X_2)) =$$

$$((cos(\psi)\tilde{X}_1 - sin(\psi)\tilde{X}_2)f)(R)$$

and finally,

$$\frac{\partial}{\partial\phi}\tilde{f}(\phi,\theta,\psi) =$$

$$f(R_z(\phi)X_3R_x(\theta).R_z(\psi)) =$$

$$f(R_z(\phi).R_x(\theta).R_z(\psi).R_z(-\psi)R_x(-\theta)X_3R_x(\theta)R_z(\psi))$$

$$= f(R.exp(-\psi.ad(X_3)).exp(-\theta.ad(X_1))(X_3))$$

Now,

$$exp(-\theta.ad(X_1))(X_3) = X_3 - \theta[X_1, X_3] + (\theta^2/2)[X_1,[X_1,X_3]] + ...$$

$$= X_2.sin(\theta) - X_3.cos(\theta)$$

$$exp(-\psi.ad(X_3)).exp(-\theta.ad(X_1))(X_3) =$$

$$exp(-\psi.ad(X_3))(X_2.sin(\theta) - X_3.cos(\theta)) =$$

$$sin(\theta).exp(-\psi.ad(X_3))(X_2) = sin(\theta)(X_2.cos(\psi) + X_1.sin(\psi))$$

Thus,

$$\frac{\partial}{\partial\phi}\tilde{f}(\phi,\theta,\psi)$$

$$= f(R.(sin(\theta)cos(\psi)X_2 + sin(\theta)sin(\psi)X_1))$$

$$= ((sin(\theta)cos(\psi)\tilde{X}_2 + sin(\theta)sin(\psi)\tilde{X}_1)f)(R)$$

Thus we obtain the following correspondences,

$$\tilde{X}_3 \to \partial/\partial\psi,$$

$$cos(\psi)\tilde{X}_1 - sin(\psi)\tilde{X}_2 \to \partial/\partial\theta$$

$$(sin(\theta)cos(\psi)\tilde{X}_2 + sin(\theta)sin(\psi)\tilde{X}_1$$

$$\to \partial/\partial\phi$$

Thus, have an expression of the form

$$\begin{pmatrix} \partial/\partial\psi \\ \partial/\partial\theta \\ \partial/\partial\phi \end{pmatrix}$$

$$= \mathbf{A}(\phi,\theta,\psi)\begin{pmatrix} \tilde{X}_1 \\ \tilde{X}_2 \\ \tilde{X}_3 \end{pmatrix}$$

where $\mathbf{A}(\phi, \theta, \psi)$ is a 3×3 matrices whose elements are functions of (ϕ, θ, ψ) and by the above discussion, it follows that the Haar density on $SO(3)$ is given in terms of Euler angles by

$$D(\phi, \theta, \psi)d\phi.d\theta.d\psi$$

where

$$D(\phi, \theta, \psi) = det(\mathbf{A}(\phi, \theta, \psi))$$

Remark: $\tilde{X}_k, k = 1, 2, 3$ are left invariant vector fields which are expressed in terms of their components w.r.t the Euler angle coordinate system as

$$\tilde{X}_k = X_{k1}(\phi, \theta, \psi)\partial/\partial\phi + X_{k2}(\phi, \theta, \psi)\partial/\theta + X_{k3}(\phi, \theta, \psi)\partial/\partial\psi, k = 1, 2, 3$$

Thus,

$$\tilde{X}_1 \wedge \tilde{X}_2 \wedge \tilde{X}_3 = det((X_{km}))\partial/\partial\phi \wedge \partial/\partial\theta \wedge \partial/\partial\psi$$

Note that

$$det((X_{km})) = 1/det(\mathbf{A})$$

if $\omega_k, k = 1, 2, 3$ is the basis of one forms on $SO(3)$ that is dual to the basis $\tilde{X}_k, k = 1, 2, 3$, then writing

$$\omega_k = \omega_{k1}d\phi + \omega_{k2}d\theta + \omega_{k3}d\psi, k = 1, 2, 3$$

we get that

$$\omega_k(\tilde{X}_m) = \delta_{km}$$

and hence since $(d\phi, d\theta, d\psi)$ is the dual basis of $(\partial/\partial\phi, \partial/\partial\theta, \partial/\partial\psi)$, it follows that

$$\omega_{kr}X_{mr} = \delta_{km}$$

where summation on the left over the repeated index r is understood. Thus,

$$\omega_1 \wedge \omega_2 \wedge \omega_3 = det((\omega_{kr}))d\phi \wedge d\theta \wedge d\psi$$

with

$$det((\omega_{kr})) = 1/det((X_{kr})) = det\mathbf{A} = D$$

[h] Induced representations for semidirect products.

[i] Semidirect products. The prototype example here is the 3-D Euclidean motion group of rotations and translations. This group is represented as

$$G = \mathbb{R}^3 \otimes_s SO(3)$$

where $\otimes_s$ denotes semidirect product and its form is derived by acting two elements g_1, g_2 of G successively on a point $x \in \mathbb{R}^3$. Let $g_1 = (a_1, R_1), g_2 =$

$(a_2, R_2) \in G$, ie, $a_1, a_2 \in \mathbb{R}^3, R_1, R_2 \in SO(3)$. Then g_1 acts on a point $x \in \mathbb{R}^3$ by rotating it by R_1 followed by translating it by a_1:

$$g_1.x = R_1 x + a_1$$

Further g_2 acts on $g_1.x$ in the same way taking it to

$$g_2.(g_1.x) = R_2(R_1 x + a_1) + a_2 = R_2 R_1 x + R_2 a_1 + a_2$$

$$= (R_2 a_1 + a_2, R_2 R_1).x$$

This formula by which the action of two successive Euclidean motion group elements acts on a point in 3-D space is used to define the composition law in G:

$$(g_2.g_1).x = g_2.(g_1.x)$$

giving the composition law in G as

$$g_2.g_1 = (a_2, R_2).(a_1, R_1) = (R_2 a_1 + a_2, R_2 R_1)$$

This composition law is easily verified to be associative. In fact, we have

$$(g_3.(g_2.g_1)).x = g_3.((g_2.g_1).x) = g_3.(g_2.(g_1.x)) = (g_3.g_2).(g_1.x)$$

$$= ((g_3.g_2).g_1).x$$

for all x by definition from which we deduce the associative property:

$$g_3.(g_2.g_1) = (g_3.g_2).g_1$$

More generally, let N be an Abelian subroup of a group G and H another subgroup of G such that

$$G = N \times_s H$$

which means that (a) Every $g \in G$ is uniquely expressible as $g = n.h$ with $n \in N, h \in H$ and (b) N is a normal subgroup of G which in view of the Abelian property of N and property (a) means that $hNh^{-1} = N \forall h \in G$. Then, we have for $g_1 = n_1 h_1, g_2 = n_2 h_2$ that

$$g_2 g_1 = n_2 h_2.n_1 h_1 = (n_2 h_2 n_1 h_2^{-1}).(h_2 h_1)$$

We can represent $g = nh \in G$ uniquely as $g = (n, h) \in N \times H$ and hence, the above composition law can also be expressed as

$$(n_2, h_2).(n_1, h_1) = (n_2 h_2 n_1 h_2^{-1}, h_2 h_1) = (n_2 \tau_{h_2}(n_1), h_2 h_1)$$

where $\tau_h(n) = hnh^{-1}$, just as in the case of the Euclidean motion group with $N = \mathbb{R}^3$ and $H = SO(3)$. More generally, let N be any Abelian group and H and other group such that there is a homomorphism $\tau : H \to aut(N)$. This means that for any $h \in H, \tau_h \in aut(N)$, ie, $\tau_h(n) \in N \forall n \in N$, $\tau_h(n_1 n_2) = \tau_h(n_1).\tau_h(n_2)$ $\forall n_1, n_2 \in N$ and $\tau_{h_2} o \tau_{h_1} = \tau_{h_2 h_1}$ for all $h_1, h_2 \in H$.

Remark: In our Euclidean motion group case, we have $\tau_R(a) = Ra, R \in SO(3), a \in \mathbb{R}^3$. All the properties required of τ are satisfied here: τ is a homomorphism from $SO(3)$ into $aut(\mathbb{R}^3)$, $aut(\mathbb{R}^3)$ being the multiplicative group of all non-singular linear transformations acting on $\mathbb{R}^3$, or equivalently, the group of all non-singular 3×3 matrices. In fact, $\tau_R = R$ here. Then, $\tau_R(a_1 + a_2) = R(a_1 + a_2) = Ra_1 + Ra_2 = \tau_R(a_1) + \tau_R(a_2)$ which proves that τ_R is an automorphism of $\mathbb{R}^3$ and secondly, $\tau_{R_2R_1}(a) = R_2R_1a = R_2.(R_1a) = \tau_{R_2}o\tau_{R_1}(a)$, proving that τ is a homomorphism of $SO(3)$ into the group $aut(\mathbb{R}^3)$. Note that the composition operation $(n_1, n_2) \to n_1n_2$ in the Abelian group $N = \mathbb{R}^3$ is here given by addition of 3-D vectors.

Coming back to the general case, we define the group $G = N \otimes_s H$ by $N \times H$ with the composition operation

$$(n_2, h_2).(n_1, h_1) = (n_2\tau_{h_2}(n_1), h_2h_1) - - - - - (a)$$

which in the Euclidean motion group case specializes to

$$(a_2, R_2).(a_1, R_1) = (a_2 + R_2a_1, R_2R_1)$$

as obtained earlier.

Exercise: Verify that (a) defines a valid associative product on G and makes G into a group.

We can show that the homomorphism τ required for defining the semidirect product is not important by proving that the same semidirect via a group isomorphism can be reduced to the standard one: $\tau_h(n) = hnh^{-1}$. Indeed, define the group $\tilde{N} = N \times \{e_H\} = \{(n, e_H) : n \in N$ and the group $\tilde{H} = \{e_N\} \times H = \{(e_N, h) : h \in H\}$. These are two subgroups of $G = N \times H$ and, let

$$(e_N, h).(n, e_H).(e_N, h)^{-1} = (n', h')$$

Then,

$$(e_N, h).(n, e_H) = (n', h').(e_N, h) = (n'\tau_{h'}(e_N), h'h) = (n', h'h)$$

or

$$(tau_h(n), h) = (n', h'h)$$

from which we get

$$h' = e_H, n' = \tau_h(n)$$

proving that

$$(e_N, h).(n, e_H).(e_N, h)^{-1} = (\tau_h(n), e_H) \in \tilde{N}$$

Thus, $\tilde{N}$ is isomorphic to N, $\tilde{H}$ is isomorphic to H and the composition in G is given by

$$(n_2, h_2).(n_1, h_1) = (n_2\tau_{h_2}(n_1), h_2h_1) =$$

$$\tilde{n}_2(\tilde{h}_2\tilde{n}_1\tilde{h}_2^{-1})\tilde{h}_2\tilde{h}_1$$

where

$$\tilde{n}_2 = (n_2, e_H), \tilde{h}_2 = (e_N, h_2),$$

and likewise for $\tilde{n}_1, \tilde{h}_1$. Thus in any semdirect product, we can always assume that the homomorphism τ from H into $aut(N)$ is always of the form $\tau_h(n) = hnh^{-1}$. For the Euclidean motion group, this fact reads as follows:

$$(0, R).(a, I).(0, R^{-1}).x = (0, R).(R^{-1}x + a) = x + Ra = (Ra, I).x$$

so that

$$(0, R).(a, I).(0, R^{-1}) = (Ra, I)$$

which is the same as saying that

$$\tau_R(a) = Ra$$

$(0, R)$ is rotation by R, $(a, 0)$ is translation by a.

[ii] Irreducible representations of a semidirect product.

Now we address the problem of determining all the inequivalent irreducible representations of a semidirect product. This will enable us to do image processing for problems involving for example, estimating both the translation and rotation vector or more generally in the case of the Galilean group of motions for images in motion, estimating the translation vector, the velocity vector, the time delay and the rotation applied to an object field defined on $\mathbb{R}^3$.

Consider first the case of a finite group G that is expressible as a semidirect product. Let N be an Abelian subgroup and H another subgroup that normalises N Suppose we can write

$$g = nh, n \in N, h \in H$$

uniquely for each $g \in G$, ie,

$$G = N \times_s H$$

Let U be an irreducible unitary representation of G in a finite dimensional Hilbert space $\mathcal{H}$. Then $U(n), n \in N$ is a commuting family of unitary operators in $\mathcal{H}$ and hence can be jointly diagonalized. This means that we can find characters $\chi_k, k = 1, 2, ..., M$ of N such that

$$U(n) = \sum_{k=1}^{M} \chi_k(n) P_k, n \in N$$

where $\{P_k : 1 \leq k \leq M\}$ is a complete spectral family in $\mathcal{H}$, ie, $P_k^* = P_k, P_k P_m = 0, k \neq m, \sum_{k=1}^{M} P_k = I$. Note that by unitarity of the $U(n)'s$ and the representation property of U, the χ_k satisfy

$$|\chi_k(n)| = 1, \chi_k(n_1 n_2) = \chi_k(n_1)\chi_k(n_2), n_1, n_2 \in N$$

Choose a $\chi_0 \in \{\chi_1, ...\chi_M\}$ and for any character χ of N, let V_χ denote the eigensubspace of $U|_N$ corresponding to the "eigenvalue" χ. In other words,

$$V_\chi = \{v \in \mathcal{H} : U(n)v = \chi(n)v\forall n \in N\}$$

For example,

$$V_{\chi_k} = P_k(\mathcal{H}) = \mathcal{R}(P_k)$$

Then, from the relation

$$U(n)U(h)v = U(h)U(h^{-1}nh)v$$

and the fact that H normalises N, it follows that

$$U(n)U(h)v = \chi_0(h^{-1}nh)U(h)v, \forall v \in V_{\chi_0}, h \in H, n \in N$$

Writing

$$\beta_h(n) = hnh^{-1},$$

and

$$\beta_h\chi(n) = \chi(\beta_h^{-1}n) = \chi(h^{-1}nh)$$

for any character χ of N, we get that

$$U(n)U(h)v = \beta_h.\chi(n)U(h)v, \forall v \in V_\chi, h \in H$$

In other words,

$$U(h)V_\chi = V_{\beta_h\chi}, h \in H, \chi \in \hat{N}$$

Here, $\hat{N}$ denotes the character group of N. In particular, we have

$$U(h)V_{\chi_0} = V_{\beta_h\chi_0}, h \in H$$

Note that

$$\beta_h\chi \in \hat{N}\forall\chi \in \hat{N}, h \in H$$

Let

$$O(\chi_0) = \{\beta_h\chi_0 : h \in H\}$$

$O(\chi_0)$ is the orbit of χ_0 in $\hat{N}$ under the action of H defined via the group action β. Note that

$$\beta_{h_2}\beta_{h_1}\chi = \beta_{h_2h_1}\chi, h_1, h_2 \in H, \chi \in \hat{N}$$

(Prove this)

We write

$$W = \bigoplus_{\chi\in O(\chi_0)} V_\chi$$

and claim that the irreducibility of U implies that

$$W = \mathcal{H}$$

In fact, this directly follows from the $U(G)$-invariance of W. To prove the $U(G)$ invariance of W, we first observe that

$$U(h)V_\chi = V_{\beta_h \chi}, h \in H, \chi \in \hat{N}$$

and $\beta_h \chi$ is in the orbit of χ which is also the orbit of χ_0 whenever χ belongs to the orbit of χ_0. Thus $U(h)$ leaves W invariant for each $h \in H$. Secondly, if $n \in N$, then

$$U(n)V_\chi = V_\chi, \chi \in \hat{N}$$

since

$$U(n)v = \chi(n)v, \forall v \in V_\chi$$

This proves that each V_χ is $U(n)$ invariant for any $\chi \in \hat{N}$ and in particular, this is true if $\chi \in O(\chi_0$. The proof that W is $U(G)$-invariant follows from this.

The next point to note that is that if we define

$$H_0 = \{h \in H : \beta_h \chi_0 = \chi_0\}$$

then H_0 is a subgroup of H, called the isotropy/little group of χ_0. We now claim that by defining

$$\sigma(h) = U(h)|_{V_{\chi_0}}, h \in H_0$$

we get a representation σ of H_0 in V_{χ_0} and further, the irreducibility of U implies the irreducibility of σ. First, note that the representation σ of H_0 is well defined since

$$U(h)V_{\chi_0} = V_{\chi_0}, h \in H_0$$

Now, suppose W_0 is a $\sigma(H_0)$-invariant subspace of V_{χ_0}. Then, for any $h \in H$, $U(h)W_0$ equals W_0 if $h \in H_0$ and otherwise, it is a subspace of $V_{\beta_h \chi_0}$. Further, $U(n)W_0 = W_0$ for each $n \in N$ since $U(n)v = \chi(n)v$ for each $v \in W_0$ because $W_0 \subset V_{\chi_0}$. Thus, we have proved that

$$W = \bigoplus_{k=1}^{r} (U(h_k)W_0)$$

is U invariant where $\{h_1, ..., h_r\} \subset H$ is any complete set of representatives of H/H_0 (ie H is the disjoint union of $h_k H_0, k = 1, 2, ..., r$) and hence, $\{\beta_{h_k}\chi_0 : k = 1, 2, ..., r\} = O(\chi_0)$. But then,

$$W = \mathcal{H}$$

since U is irreducible. Hence it must be true that $W_0 = V_{\chi_0}$ (Note that the subspaces $U(h_k)W_0, k = 1, 2, ..., r$ are mutually orthogonal subspaces of $\mathcal{H}$, each having the same dimension, $dim W_0$. They are orthogonal because $U(h_k)W_0 \subset V_{\beta_{h_k}\chi_0}$ and the latter are all orthogonal because they are the eigensubspaces of $U|_N$ with different eigenvalues and $U|_N$ is a unitary representation). This completes the proof of the irreduciblity of σ as a unitary representation of the little group.

Conversely, suppose σ is any irreducible representation of the little group of a character $\chi_0 \in \hat{N}$ (appearing in the semidirect product $G = N \times_s H$) in the Hilbert space V_0. Then we can reverse the entire argument above to arrive at an irreducible representation U of G in a Hilbert space $\mathcal{H}$. Formally, to see how this construction is carried out, we first construct the H orbit of χ_0:

$$O(\chi_0) = \{\beta_h \chi_0 : h \in H\} = \{\beta_{h_k} \chi_0, k = 0, 1, 2, ..., r-1\}$$

where $h_k H_0, k = 0, 1, 2, ..., r-1$ with $h_1 = e$ are all the distinct and hence disjoint set of cosets of H_0 in H:

$$H = \bigcup_{k=0}^{r-1} h_k H_0$$

We then formally attach a vector space V_k to the character $\beta_{h_k} \chi_0$ for each $k = 0, 1, ..., r-1$ so that

$$U(h_k) V_0 = V_k, k = 0, 1, .., r-1$$

Then define

$$\mathcal{H} = \bigoplus_{k=0}^{r-1} V_k$$

as an orthogonal direct sum. It remains to define the action of $U(G)$ on $\mathcal{H}$ compatible with the above definitions. This is done as follows. Let $g = nh \in G, n \in N, h \in H$. Then let $v \in V_k$ for some $k = 0, 1, 2, ..., r-1$. Let $s \in \{0, 1, ..., r-1\}$ be the unique element such that

$$hh_k \in h_s H_0$$

Then,

$$U(h)v \in V_s$$

will be defined as follows. Choose an onb $\{\phi_{0,1}, ..., \phi_{0,m}\}$ for V_0 and then for each $k = 1, 2, ..., r-1$, define an onb $\{\phi_{k,1}, ..., \phi_{k,m}\}$ for V_k so that

$$U(h_k)\phi_{0,l} = \phi_{k,l}, k = 0, 1, ..., r-1, l = 1, 2, ..., m$$

Then, for $h \in H$, we can write $hh_k \in h_s H_0$ as above. Thus, $hh_k = h_s h_0'$ for some $h_0' \in H_0$. Then

$$U(h)\phi_{k,l} = U(h)U(h_k)\phi_{0,l} = U(hh_k)\phi_{0,l}$$

$$= U(h_s h_0')\phi_{0,l} = U(h_s)U(h_0')\phi_{0,l} =$$

$$U(h_s)\sigma(h_0')\phi_{0,l} = U(h_s) \sum_{l'=1}^{m} [\sigma(h_0')]_{l'l} \phi_{0,l'}$$

$$= \sum_{l'=1}^{m} [\sigma(h_0')]_{l'l} U(h_s)\phi_{0,l'}$$

$$= \sum_{l'=1}^{m} [\sigma(h_0')]_{l'l} \phi_{s,l'}$$

This formula defines the representation U of G on the space $\mathcal{H}$ spanned by the onb $\{\phi_{k,l} : 0 \leq k \leq r-1, 1 \leq l \leq m\}$. Note how the action of $U(n), n \in N$ is defined:

$$U(n)\phi_{k,l} = \beta_{h_k} \chi_0(n)\phi_{k,l} = \chi_0(h_k^{-1} n h_k)\phi_{k,l}$$

Note that this implies

$$U(n)V_k = V_k, k = 0, 1, ..., r-1$$

By reversing the argument above, it is easily proved that U is an irreducible unitary representation of G in $\mathcal{H}$. In fact, we first observe that since $\{\phi_{k,l} : 1 \leq l \leq m\}$ is an onb for V_k by definition, and $U(h_k)\phi_{0,l} = \phi_{k,l}$, it follows immediately that

$$U(h_k)V_0 = V_k, k = 0, 1, ..., r-1$$

Now suppose that W is a $U(G)$-invariant subspace of $\mathcal{H}$. Then consider the subspace $W_0 = W \cap V_0$ of V_0. We claim that W_0 is a $\sigma(H_0)$-invariant subspace of V_0. This immediately follows from the definitions. Hence by the irreducibility of $\sigma(H_0)$, it follows that $W_0 = V_0$ and therefore that $V_0 \subset W$ and since

$$U(h_k)V_0 = V_k, k = 0, 1, ..., r-1$$

and

$$U(h_k)W = W, k = 0, 1, ..., r-1$$

by W-invariance of $U(G)$, it follows that

$$V_k = U(h_k)V_0 \subset U(h_k)W = W, k = 0, 1, ..., r-1$$

Therefore

$$\mathcal{H} = \bigoplus_{k=0}^{r-1} V_k \subset W$$

ie,

$$W = \mathcal{H}$$

proving the irreduciblity of $U(G)$. This completes the construction. Note that the dimension of the irreducible representation U of G is related to the dimension of the irreducible representation σ of the little group H_0 by

$$dimU = rm = o(H/H_0).dim(\sigma) = o(O(\chi_0)).dim(\sigma)$$

Let G be a compact group acting transitively on a manifold $\mathcal{M}$. The prototype example of this is $SO(3)$ acting on S^2. Take an image field $f_1 : \mathcal{M}] \to \mathbb{C}$.

The image field f_2 on $\mathcal{M}$ after transforming it by a $g \in G$ and adding noise to it is given by the statistical model

$$f_2(x) = f_1(g^{-1}x) + w(x), x \in \mathcal{M}$$

[i] The irreducible representations of S_n-the permutation/symmetric group. Preliminaries:

[1] The group algebra of a finite group. let G be a finite group. Its group algebra consists of all formal linear combinations

$$f = \sum_{g \in G} f(g)g$$

where $f : G \to \mathbb{C}$ is arbitrary. We denote this set by $\mathcal{A}(G)$. If $f_1, f_2 \in \mathcal{A}(G)$, their product is defined by

$$f_1 f_2 = \sum_{g,h \in G} f_1(g) f_2(h) gh = \sum_{g \in G} (f_1 * f_2)(g) g$$

where

$$(f_1 * f_2)(g) = \sum_{h \in G} f_1(gh^{-1}) f_2(h) = \sum_{h \in G} f_1(h) f_2(h^{-1}g)$$

is called the convolution of the functions f_1 and f_2 on G. Addition and scalar multiplication in $\mathfrak{A}(G)$ are defined in the usual way:

$$cf_1 + f_2 = \sum_{g \in G} (cf_1(g) + f_2(g))g, c \in \mathbb{C}$$

With these operations, $\mathcal{A}(G)$ becomes an algebra and is called the group-algebra of G. Equivalently, $\mathcal{A}(G)$ can be viewed as the set of all complex valued functions on G with multiplication defined by the convolution operation as above and addition and scalar multiplication defined in the usual way, ie, pointwise on G.

[2] Minimal projections. G is again assumed to be a finite group and $\hat{G}$ the set of all inequivalent irreducible unitary representations of G. Note that since G is finite, any finite dimensional representation of G is equivalent to a unitary representation. We denote by $\{D_\alpha(g) : \alpha \in \hat{G}\}$ a complete set of irreducible unitary representations of G. We have by the Peter-Weyl theorem, for any function f on G,

$$f(g) = \sum_{1 \le i,j \le d(\alpha), \alpha \in G} c(\alpha, i, j)[D_\alpha]_{ij}(g)$$

where $d(\alpha)$ is the dimension of the representation D_α and

$$c(\alpha, i, j) = d(\alpha) \sum_{g \in G} f(g)[\bar{D}_\alpha(g)]_{ij}$$

A projection $p \in \mathfrak{A}(G)$ is defined by the condition

$$p^2 = p$$

Let p be a projection. Then, by Peter-Weyl theorem,

$$p = \sum_{g \in G} p(g)g = \sum_{\alpha \in \hat{G}} p_\alpha$$

where

$$p_\alpha(g) = d(\alpha) \sum_{i,j=1}^{d(\alpha)} < p, [D_\alpha]_{ij} > [D_\alpha(g)]_{ij}$$

or equivalently,

$$p_\alpha = d(\alpha) \sum_{i,j} < p, [D_\alpha]_{ij} > [D_\alpha]_{ij}$$

where

$$[D_\alpha]_{ij} = \sum_{g \in G} [D_\alpha(g)]_{ij} . g$$

and

$$< u, v >= o(G)^{-1} \sum_{g \in G} u(g)\bar{v}(g)$$

for any two functions u, v on G. Now, by the Schur orthogonality relations, we have as elements of $\mathcal{A}(G)$, and with

$$c(p, \alpha, i, j >= d(\alpha) < p, [D_\alpha]_{ij} >$$

$$p_\alpha . p_\beta = \sum_{g \in G} (p_\alpha * p_\beta)(g) . g$$

and

$$(p_\alpha * p_\beta)(g) = \sum c(p, \alpha, ij) c(p, \beta, km)([D_\alpha]_{ij} * [D_\beta]_{km})(g)$$

with

$$([D_\alpha]_{ij} * [D_\beta]_{km})(g) =$$

$$\sum_{h \in G} [D_\alpha(h)]_{ij} . [D_\beta]_{km}(h^{-1}g)$$

$$= \sum_{h \in G} [D_\alpha(h)]_{ij} [D_\beta(h^{-1})]_{kl} [D_\beta(g)]_{lm}$$

$$= \sum_{h \in G} [D_\alpha(h)]_{ij} [] bar D_\beta(h)]_{lk} [D_\beta(g)]_{lm}$$

and this is zero if $\alpha \neq \beta$ (Schur's orthogonality relation which states that matrix elements of inequivalent irreducible unitary representations are orthogonal) and if $\alpha = \beta$, then this is contained in the vector space of functions

$$V_\alpha = span\{[D_\alpha(g)]_{ij} : 1 \leq i, j \leq d(\alpha)\}$$

Thus we get in $\mathfrak{A}(G)$,

$$p_\alpha . p_\beta = 0, \alpha \neq \beta, p_\alpha^2 \in V_\alpha$$

Hence,

$$\sum_\alpha p_\alpha p = p^2 = \sum_{\alpha,\beta} p_\alpha p_\beta = \sum_\alpha p_\alpha^2$$

which implies (since p_α, p_α^2 are in V_α and the $V_\alpha' s$ are mutually all orthogonal that

$$p_\alpha^2 = p_\alpha$$

or equivalently in terms of functions,

$$(p_\alpha * p_\beta)(g) = \delta(\alpha, \beta) p_\alpha, \alpha, \beta \in \hat{G}$$

A projection p is said to be minimal if it cannot be decomposed as

$$p = p_1 + p_2$$

with both p_1 and p_2 being projections. Thus, by the above decomposition, p is minimal iff $p = p_\alpha$ for some $\alpha \in \hat{G}$. We have thus proved that the set of all minimal projections in the group algebra of a finite group is in one-one correspondence with the set of all irreducible representations of G. Now suppose p is a minimal projection associated to $\alpha \in \hat{G}$. Then, we can write

$$p(g) = \sum_{i,j=1}^{d(\alpha)} c(i,j)[D_\alpha(g)]_{ij}$$

or equivalently,

$$p = \sum_{ij} c(ij)[D_\alpha]_{ij}$$

in the group algebra $\mathcal{A}(G)$. The condition $p^2 = p$ implies that

$$\sum_{im,g} c(im)[D_\alpha(g)]_{im} . g = p = \sum_{ijkm} c(ij)c(km)[D_\alpha]_{ij} . [D_\alpha]_{km}$$

$$= \sum_{ijkm,g} c(ij)c(km)[D_\alpha]_{ij} * [D_\alpha]_{km}(g) g$$

$$= \sum_{ijkm,l} c(ij)c(km)(\sum_{h \in G} [D_\alpha(h)]_{ij} . [\bar{D}_\alpha(h)]_{lk}) \sum_{g \in G} [D_\alpha(g)]_{lm}(g) . g$$

$$= \sum_{ijkml} c(ij)c(km) o(G) . d(\alpha)^{-1} \delta(i,l) \delta(j,k) . \sum_{g \in G} [D_\alpha(g)]_{lm} . g$$

$$= \sum c(ij)c(jm) . o(G) . d(\alpha)^{-1} [D_\alpha(g)]_{im} g$$

and therefore,

$$c(im) = \sum_j c(ij)c(jm) o(G) . d(\alpha)^{-1}$$

or equivalently, in matrix notation,

$$(d(\alpha)/o(G))\mathbf{C} = \mathbf{C}^2$$

If we further impose the restriction that p is a central projection, ie, p commutes with $\mathcal{A}(G)$ apart from being minimal, then the only solution to the above equation is

$$\mathbf{C} = (d(\alpha)/o(G))\mathbf{I}_{d(\alpha)}$$

and hence we find that in this case

$$p(g) = (d(\alpha)/o(G))\sum_{i=1}^{d(\alpha)}[D_\alpha(g)]_{ii} = (d(\alpha)/o(G))\chi_\alpha(g)$$

where $\chi_\alpha(g) = Tr(D_\alpha(g))$ is the character of the representation D_α.

Prove that in $\mathcal{A}(G)$,

$$p = \sum_{i,j=1}^{d(\alpha)} c(ij)[D_\alpha]_{ij}$$

commutes with all $[D_\beta]_{ij}, i,j = 1,2,...,d(\beta), \beta \in \hat{G}, \beta \neq \alpha$. (In fact, the Schur orthogonality relations imply that $p.[D_\beta]_{ij} = [D_\beta]_{ij}p = 0, \beta \neq \alpha$. Prove further that p also commutes with $[D_\alpha]_{ij}, i,j = 1,2,...,d(\alpha)$ iff $\mathbf{C} = ((c(ij)))$ is a scalar multiple of $\mathbf{I}_{d(\alpha)}$.

Remark: We have shown that the problem of determining all the minimal central projections in $\mathcal{A}(G)$ is equivalent to determining all the irreducible characters of G which is in turn equivalent to determining all the inequivalent irreducible representations of G. This fact will play a fundamental role in our determination of all the irreducible representations of the permutation groups.

[3] S_m is the group of permutation of m elements. Any $\sigma \in S_m$ can be represented as

$$\sigma = (i_1,...,i_{l_1}).(i_{l_1+1},...,i_{l_1+l_2})...(i_{l_1+...+l_{k-1}}+1,...,i_{l_1+...+l_k})$$

where $(i_1,...,i_{l_1+...+l_k})$ is a permutation of $(1,2,...,m)$ and if $a_1,...,a_r$ are distinct integers in $\{1,2,..,m\}$, then $(a_1,...,a_r)$ denotes the cyclic permutation that sends $a_i \to a_{i+1}, i = 1,2,...,r-1, a_r \to a_1$ and leaves the other integers fixed. In short, $(a_1,...,a_r)$ is a cyclic permutation in S_m with cycle length r. We thus state this result as: Every permutation is a product of disjoint cycles. Further, in the above notation, let ρ denote the permutation $\{1,2,...,m\} \to \{i_1,...,i_m\}$ where of course $m = l_1 + ... + l_k$. We also define the permutation

$$g = (1,2,...,l_1).(l_1+1,...,l_1+l_2)...(l_1+...+l_{k-1}+1,...,l_1+...+l_k)$$

expressed as a product of cycles. Then, it is clear that

$$\sigma.\rho = \rho.g$$

ie,

$$\sigma = \rho.g.\rho^{-1}$$

It is clear from this formula, that each conjugacy class in S_n consists precisely of those elements having the same cycle structure. More precisely we say that a permutation $\sigma \in 1^{k_1}2^{k_2}..m^{k_m}$ iff in the cycle representation of σ, there are k_j cycles of length j for each $j = 1, 2, ..., m$. Of course we must have $\sum_{j=1}^{m} j.k_j = m$. The conjugacy classes in S_m are therefore labeled by the integers $(k_1, ..., k_m)$. $1^{k_1}..m^{k_m}$ is a conjugacy class. The number of elements in this conjugacy class is easily seen to be

$$\mu(k_1, ..., k_m) = \frac{m!}{k_1!...k_m!1^{k_1}...m^{k_m}}$$

in fact, first simply write down all the cycles in this class serially as above in non-decreasing order of their lengths. Then we can permute all the m elements in this serial representation in $m!$ ways. However, a given cycle of length j can be represented in j possible ways by simply by applying a cyclic permutation to the elements in this cycle and there are j possible cyclic permutations. So the number of permutations within each cycle which do not alter the cyclic representation is $\Pi_j j^{k_j}$ because there are k_j cycles of length j and each such cycle can be represented in j possible ways by applying cyclic permutations. Further, the cyclic representation of a permutation is not altered if we simply permute the cycles of the same length amongst themselves. The total number of such permutations is simply $k_1!...k_m!$. Thus we obtain the above formula.

[4] Young frames and Young Tableaux: Given positive integers $m_1 \geq m_2 \geq ... \geq m_k > 0$, such that $m_1 + ... + m_k = m$, we draw a tableaux consisting of rows of boxes one below the other starting at the same left end line such that the first row has m_1 boxes, the second row has m_2 boxes,... the kth row has m_k boxes. We denote such a frame by $F(m_1, ..., m_k)$ and call it a Young frame. It is easily seen that the total number of Young frames for fixed m is simply the total number of conjugacy classes of S_m. In fact, we can directly construct a bijection from the set of all conjugacy classes onto the set of all Young frames as follows. Given a conjugacy class $1^{k_1}..m^{k_m}$, some of the j^{k_j} will be missing, ie, $k_j = 0$ for such j. Thus, we can express this conjugacy class as $r_1^{l_1}...r_p^{l_p}$ where $l_1, ..., l_p > 0$ and $1 \leq r_1 < r_2 < ... < r_p \leq m$, ie, in each element of this conjugacy class, there are l_j cycles of length j for each $j = 1, 2, ..., p$. Then we define $m_1 = ... = m_{l_p} = r_p, m_{l_p+1} = .. = m_{l_p+l_{p-1}} = r_{p-1}$ etc, ie, in other words, as an ordered k-tuple,

$$(m_1, m_2, ..., m_k) = (r_p, .., r_p, r_{p-1}, ..., r_{p-1}, ..., r_1, ..., r_1)$$

where r_j occurs l_j times for each $j = p, p-1, ..., 1$. Note that

$$m = \sum_{j=1}^{m} jk_j = \sum_{j=1}^{p} l_j r_j = \sum_{j=1}^{k} m_j$$

Since the number of distinct classes of $G = S_n$ equals the total number of inequivalent irreducible representations of G, it follows that this number is also the same as the number of distinct Young frames.

A Young tableau $F(T)$ corresponding to a Young frame $F = F(m_1, ..., m_k)$ is simply an arrangement of the m integers $1, 2, ..., m$ in these $m = \sum_{j=1}^{k}$ boxes, ie, in each box, we put an integer from $1, 2, ..., m$ and no two boxes have the same integer. Associated with the Young tableaux $F(T)$, we define $R(T)$ to be the group of all permutations of each row of $F(T)$ and $C(T)$ to be the group of all permutations of each column of $F(T)$. Note that the number of elements in $R(T)$ equals $m_1!...m_k!$ and the number of elements in $C(T)$ equals $n_1!...n_r!$ where $n_1, ..., n_r$ are the column lengths of $F(T)$. Note that $R(T)$ and $C(T)$ are subgroups of S_m. Define

$$E(T) = \sum_{p \in R(T), q \in R(T)} pq(-1)^q = P(T)Q(T)$$

where

$$P(T) = \sum_{p \in R(T)} p, Q(T) = \sum_{q \in C(T)} (-1)^q q$$

These elements are understood to be interpreted as elements of the group algebra $A(S_m)$ of S_n. Our aim is to prove that corresponding to each Young frame F, there is exactly one irreducible character χ_F of S_m, or equivalently, exactly one inequivalent irreducible representation of S_m and as F runs over all the Young frames, χ_F will run over all the distinct irreducible characters of S_m. The second part is obvious if we can prove that for two distinct Young frames F, F', the irreducible characters $\chi_F, \chi_{F'}$ associated with them are distinct since as remarked above, the number of distinct Young frames equals the number of classes of S_m and the number of classes of finite group equals exactly the number of inequivalent irreducible representations of the group.

Now let T, T' be two Young tableaux. We say that TRT', ie, T is related to T' if there exists a pair of indices (i, j) belonging to the column of T and to the same row of T', otherwise, we say that $TNRT'$, ie, T is not related to T'. now suppose TRT'. Then we have for some (i, j) that

$$(i, j) \in C(T) \cap R(T')$$

and therefore, with (i, j) denoting the transposition of (i, j), we have

$$(i, j)P(T') = P(T'), C(T)(i, j) = -C(T)$$

Now let T, T' be two Tableaux with corresponding frames $F = F(T), F' = F(T')$. Note that $F = F(T)$ is completely specified by integers $m_1 \geq m_2 \geq ... \geq m_k > 0$ and $F' = F(T')$ is completely specified by integers $m'_1 \geq m'_2 \geq ... \geq m'_l > 0$. We say that $F = F'$ if $l = k$ and $m_j = m'_j, j = 1, 2, ..., k$. Otherwise, we write $F \neq F'$. We write $F > F'$ if for the first index $j = 1, 2, ...$ for which $m_j \neq m'_j$, we have $m_j > m'_j$. Obviously $F \neq F'$ iff either $F > F'$ or else $F' > F$. Now suppose $F = F'$ and $TNRT'$. First observe that by definition of NR, all the entries in the first column of T occupy different rows of T'. Then consider the first two entries in the first column of T. By definition of NR, these two entries must fall in different rows of T'. Hence, by applying a column permutation q_1 to T and a row permutation p_1' to T', we can ensure that these two entries occupy the same positions in $q_1 T$ and in $p_1' T'$. Then we also observe that $q_1 T N R p_1' T'$ and hence by applying the same argument to these new Tableaux, we can ensure the existence of a column permutation q_2 of $q_1 T$ and a row permutation p_2' of $p_1' T'$ such that that the next pair of entries in the first column $q_2 q_1 T$ occupy the same positions in $p_2' p_1' T'$ without altering the positions of the previous pair in the two tableaux. In this way, we finally end up with a column permutation q of T and a row permutation p' of T' such that $qT = p'T'$. Then, $T' = p'^{-1} qT$ and we get

$$T = q^{-1} p' T'$$

and

$$q^{-1} p' q \in q^{-1} R(T') q = q^{-1} R(p'T') q = R(q^{-1} p' T') = R(T)$$

Thus defining

$$p = q^{-1} p' q$$

we get that

$$p \in R(T), q \in C(T), q^{-1} p' = p q^{-1}, T = p q^{-1} T', T' = q p^{-1} T$$

We have thus proved the following theorem:

Theorem: Let T, T' have the same shape. Then, if $TNRT'$, there exists a $p \in R(T), q \in C(T)$ such that $T' = qpT$.

Now, let T be any tableaux and let $g \in S_m$ be such that $g \notin R(T).C(T)$. Then, $g^{-1} \notin C(T)R(T)$ and we prove the existence of $p_0 \in R(T), q_0 \in C(T)$ such that $(-1)^{q_0} = -1, p_0 g q_0 = g$. Indeed, define $T' = g^{-1} T$. Then T' cannot be written as abT where $a \in C(T), b \in R(T)$. In fact the equation $T' = hT$ uniquely determines h as g^{-1}. Thus, by the previous theorem, TRT' and hence there exists a pair (i, j) falling in the same column of T and the same row of T'. Define

$$q_0 = (i, j), p_0 = g q_0^{-1} g^{-1} = g q_0 g^{-1}$$

Then, we get

$$(-1)^{q_0} = -1, p_0 g q_0 = g,$$

and

$$p_0 = g q_0 g^{-1} \in g R(T') g^{-1} = R(gT') = R(T)$$

and the proof of the claim is complete. We state this as a theorem.

Theorem: If $g \notin R(T)C(T)$, then there exists a $p_0 \in R(T)$ and a $q_0 \in C(T)$ such that $(-1)^{q_0} = -1 and p_0 g q_0 = g$.

Now we are in a position to prove one of the main theorems in the Frobenius-Young theory.

Theorem: Let $s \in (S_n)$ be such that

$$psq = (-1)^q s \forall p \in P(T), q \in Q(T)$$

Then

$$s = s(e)E(T)$$

Proof: Write

$$s = \sum_{g \in G} s(g)g, G = S_n$$

Then the stated hypothesis implies that

$$s(pgq) = (-1)^q s(g), \forall g \in G, p \in R(T), q \in C(T)$$

(Note that $R(T), C(T)$ are subgroups of G so that $p \in R(T)$ iff $p^{-1} \in P(T)$ and likewise $q \in C(T)$ iff $q^{-1} \in C(T)$). Taking $g = e$ gives us

$$s(pq) = (-1)^q s(e), p \in P(T), q \in Q(T)$$

Now suppose $g \notin P(T)Q(T)$. Then by the previous theorem, there exist $p_0 \in P(T), q_0 \in Q(T)$ such that

$$(-1)^{q_0} = -1, p_0 g q_0 = g$$

Then,

$$s(g) = s(p_0 g q_0) = (-1)^{q_0} s(g) = -s(g)$$

and hence $s(g) = 0$. Therefore, we have proved that

$$s = \sum_{p \in P(T), q \in Q(T)} s(pq)pq = s(e) \sum_{p \in P(T), q \in Q(T)} (-1)^q pq = s(e)E(T)$$

and the proof of the theorem is complete.

Corollary: $E(T)^2 = k(T)E(T)$ for some $k(T) \in \mathbb{R}$. In fact, we have

$$E(T) = P(T)Q(T), E(T)^2 = P(T)Q(T)P(T)Q(T)$$

and hence,

$$pE(T)^2 q = pP(T)Q(T)P(T)Q(T)q = (-1)^q E(T)^2, p \in P(T), q \in Q(T)$$

since

$$pP(T) = P(T), Q(T)q = (-1)^q Q(T), p \in P(T), q \in Q(T)$$

Thus, by the theorem,

$$E(T)^2 = k(T)E(T)$$

for some $k(T) \in \mathbb{R}$. To evaluate the value of $k(T)$, we evaluate $Tr(R_{E(T)})$ and $Tr(R_{E(T)}^2) = Tr(R_{E(T)}^2)$, where for any $s \in A(G)$, we define the linear operator on the vector space $A(G)$ by

$$R_s f = fs$$

Then, for any $g \in G$, it is clear by choosing the standard basis $\{h : h \in G\}$ for $A(G)$ that

$$Tr(R_g) = 0, g \notin e, Tr(R_e) = m!$$

(Recall that $G = S_m$). Hence, since

$$R_s = \sum_{g \in G} s(g) R_g, s \in A(G)$$

we get that

$$R_{E(T)} = \sum_{p \in P(T), q \in Q(T)} (-1)^q R_{pq}$$

so that

$$Tr(R_{E(T)}) = m!$$

Now let

$$d = rank(R_{E(T)}) = dim\mathcal{R}(R_{E(T)}) = dim[A(G)E(T)]$$

Then, let $\{f_1, ..., f_d\}$ be a basis for $(R_{E(T)})$. We define the linear operator

$$A = k(T)^{-1} R_{E(T)}$$

acting on the vector space $A(G)$. Then,

$$A^2 = A$$

ie, A is a projection. Thus,

$$Tr(A^2) = Tr(A) = d$$

Combining these two formulae gives us

$$d = Tr(k(T)^{-1} R_{E(T)}) = k(T)^{-1} m!$$

so that

$$k(T) = m!/d$$

and we conclude that

$$E(T)^2 = (m!/d)E(T)$$

Now define

$$e(T) = (d/m!)E(T)$$

Then,

$$e(T)^2 = (d/m!)^2 E(T)^2 = (d/m!)E(T) = e(T)$$

in other words, $e(T)$ is an idempotent element of the group algebra $A(G)$. We also note that if T, T' are different Tableaux having the same shape $F = F(T) = F(T') = F'$, then $e(T)e(T') = 0$.

[j] The Frobenius character formula for induced representations of a finite group.

[1] Alternate definition of the induced representation.

Let G be a finite group and H a subgroup of G. Let L be a unitary representation of H in the Hilbert space Y. We define

$$U = Ind_H^G L$$

ie, U is the representation of G induced by the representation L of H. There are many equivalent ways to define U. All these definitions give isomorphic representations of G. One way is define the representation space X of U as the set of all $f \in C(G, Y)$ for which $f(gh) = L(h)^{-1}f(g), h \in H, g \in G$ and then $U(g)f(x) = f(g^{-1}x), g, x \in G, f \in X$. It should be noted from this definition that $f \in X$ is completely determined if its value is known on any set of representatives of the cosets G/H. Hence, we may equivalently view any $f \in X$ as a map from $G/H \to Y$. In fact, let γ be a cross section map for this coset space, ie, $\gamma : G/H \to G$ is such that $\gamma(x)H = x, x \in G/H$. Then consider the element of G

$$h(g, x) = \gamma)(x)^{-1}g\gamma(g^{-1}x), g \in G, x \in G/H$$

Clearly since

$$h(g, x)H = \gamma(x)^{-1}g\gamma(g^{-1}x)H = \gamma(x)^{-1}gg^{-1}xH = H$$

it follows that

$$h(g, x) \in H, g \in G, h \in H$$

Then for a mapping $\psi : G/H \to Y$, consider

$$(V(g)\psi)(x) = L(h(g, x))\psi(g^{-1}x)$$

We observe that

$$(V(g_2)(V(g_1)\psi))(x) = L(h(g_2, x))(V(g_1)\psi)(g_2^{-1}x)$$

$$= L(h(g_2, x))L(h(g_1, g_2^{-1}x))\psi(g_1^{-1}g_2^{-1}x)$$

$$= L(h(g_2, x).h(g_1, g_2^{-1}x))\psi((g_2g_1)^{-1}x)$$

and

$$h(g_2, x))h(g_1, g_2^{-1}x) =$$

$$(\gamma(x)^{-1}g_2\gamma(g_2^{-1}x)).(\gamma(g_2^{-1}x)^{-1}g_1\gamma(g_1^{-1}g_2^{-1}x)$$

$$= \gamma(x)^{-1}g_2g_1\gamma((g_2g_1)^{-1}x) = h(g_2g_1, x)$$

and therefore,

$$V(g_2)V(g_1)\psi(x) = L(g_2g_1, x)\psi((g_2g_1)^{-1}x) = V(g_2g_1)\psi(x),$$

ie, as operators in $C(G/H, Y)$, the set $\{V(g) : g \in G\}$, satisfies

$$V(g_2)V(g_1) = V(g_2g_1), g_1, g_2 \in G$$

Thus, $V(.)$ is a representation of G in $C(G/H, Y)$. We wish to prove that V is equivalent to the induced representation $U = Ind_H^G L$. Indeed, consider the map

$$T : C(G/H, Y) \to X$$

defined by

$$(T\psi)(g) = L(h(g^{-1}, H))\psi(gH), g \in G$$

To see that this map is well defined, we must first show that the rhs is indeed an element of X, ie, it satisfies

$$(T\psi)(gh) = L(h)^{-1}(T\psi)(g), g \in G, h \in H$$

Indeed, this follows from

$$(T\psi)(gh) = L(h((gh)^{-1}, H))\psi(gH)$$

and

$$h((gh)^{-1}, H) = \gamma(H)^{-1}(gh)^{-1}\gamma(ghH) = (gh)^{-1}\gamma(gH)$$

on assuming without loss of generality,

$$\gamma(H) = e$$

Thus,

$$h((gh)^{-1}, H) = h^{-1}h(g^{-1}, H)$$

and therefore,

$$(T\psi)(gh) = L(h)^{-1}L(h(g^{-1}, H)\psi(gH) = L(h)^{-1}(T\psi)(g)$$

proving thereby that

$$T\psi \in X$$

and hence the map T is well defined. We next show that T is a bijection. Indeed, $T\psi = 0$ implies

$$L(h(g^{-1}, H))\psi(gH) = 0, g \in G$$

which implies that

$$\psi(gH) = 0, g \in G$$

ie,

$$\psi = 0$$

Thus, T is injective. Next, suppose $\phi \in X$. Then, define $\psi \in C(G/H, Y)$ by

$$\psi(gH) = L(h(g^{-1}, H))^{-1}\phi(g), g \in G$$

To show that ψ is a well defined element of $C(G/H, Y)$, we must show that

$$L(h((gh)^{-1}, H))^{-1}\phi(gh) = L(g^{-1}, H))^{-1}\phi(g), g \in G, h \in H$$

But, for all $h \in H, g \in G$

$$h((gh)^{-1}, H) = \gamma(H)^{-1}h^{-1}g^{-1}\gamma(ghH)$$

$$= h^{-1}g^{-1}\gamma(gH) = h^{-1}.h(g^{-1}, H)$$

Then,

$$L(h((gh)^{-1}, H))^{-1}\phi(gh) = L(h^{-1}h(g^{-1}, H))^{-1}.L(h)^{-1}\phi(g)$$

$$= L(h(g^{-1}, H))^{-1}\phi(g)$$

proving that $\psi \in C(G/H, Y)$ and by construction, $T\psi = \phi$. Thus, T is also surjective. Hence, T is bijective. Finally, we prove that T intertwines the representations U and V and this will establish the equivalence of these two representations and hence provide an alternate equivalent definition of the induced representation. We have for $\psi \in C(G/H, Y)$,

$$(U(g)T\psi)(g_1) = (T\psi)(g^{-1}g_1) =$$

$$L(h(g_1^{-1}g, H))\psi(g^{-1}g_1H)$$

on the one hand, while on the other

$$(TV(g)\psi)(g_1) = L(h(g_1^{-1}, H))(V(g)\psi)(g_1H)$$

$$= L(h(g_1^{-1}, H)).L(h(g, g_1H))\psi(g^{-1}g_1H)$$

$$= L(h(g_1^{-1}, H).h(g, g_1H))\psi(g^{-1}g_1H)$$

and since

$$h(g_1^{-1}, H).h(g, g_1H) = \gamma(H)^{-1}g_1^{-1}\gamma(g_1H).\gamma(g_1H)^{-1}g\gamma(g^{-1}g_1H)$$

$$= g_1^{-1}g\gamma(g^{-1}g_1H) = h(g_1^{-1}g, H)$$

it follows that

$$U(g)T = TV(g), g \in G$$

ie, T intertwines the represenations U of G in X and V of G in $C(G/H, Y)$ and since T is a bijection, we can write this intertwining relation as

$$V(g) = T^{-1}U(g)T, g \in G$$

proving the equivalence of U and V.

[2] Frobenius character formula for the induced representation.

Let G be a finite group and H a subgroup of G. Let L be a unitary representation of H and let $G = Ind_H^G L$ be the induced representation. Let L act in the Hilbert space Y. Then Let $\{phi_1, \ldots, \phi_n\}$ be an onb for Y. Define $c = \sqrt{o(G)/o(H)}$ and $f_k(g) = c.L(g)^{-1}\phi_k, ifg \in H$ and $f_k(g) = 0$ if $g \in G - H$. We claim that $\{f_1, \ldots, f_n\}$ is an on set in the representation space X of U. Note that this space X is defined as the set of all elements $f \in C(G, Y)$ for which $f(gh) = L(h)^{-1}f(g), g \in G, h \in H$. First we observe that if $g \in G - H, h \in H$, then $gh \in G - H$ in which case $f_k(gh) = 0 = f_k(g)$ by definition. Thus in this case, $f_k(gh) = L(h)^{-1}f_k(g) = -0$ holds. Next we observe that if $g, h \in H$, then $gh \in H$ in which case, we get by definition, $f_k(gh) = c.L(gh)^{-1}\phi_k = L(h)^{-1}c.L(g)^{-1}\phi_k = L(h)^{-1}f_k(g)$. This completes the proof of the claim that $f_k \in X, k = 1, 2, \ldots, n$. Next we calculate

$$< f_k, f_m >= \frac{1}{o(G)} \sum_{g \in G} < f_k(g), f_m(g) >$$

$$= \frac{c^2}{o(G)} \sum_{h \in H} < L(h)^{-1}\phi_k, L(h)^{-1}\phi_m >$$

$$= \frac{c^2}{o(G)} \sum_{h \in H} < \phi_k, \phi_m >= \frac{c^2 o(H)}{o(G)} \delta_{km} = \delta_{km}$$

Since

$$c = \sqrt{o(G)/o(H)}$$

Thus, $\{f_1, \ldots, f_n\}$ is an onb for X. Let $\{k_1 = e, k_2, , \ldots, k_r\}$, be a complete set of representatives of G/H, ie, $k_j H, j = 1, 2, .., r$ are all disjoint elements in G/H and $\bigcup_{j=1}^r k_j H = G$. Then we claim that $\mathcal{B} = \{U(k_j)f_k, j = 1, 2, \ldots, r, k = 1, 2, \ldots, n\}$ is an onb for X. First observe that

$$dimX = (o(G)/o(H)).dimY = rn$$

So if we are able to prove that the elements of $\mathcal{B}$ defined above are orthonormal, then our claim will be proved. But,

$$< U(k_j)f_k, U(k_l)f_m >= o(G)^{-1}. \sum_{g \in G} < f_k(k_j^{-1}g), f_m(k_l^{-1}g) >=$$

$$= o(G)^{-1}. \sum_{g \in G} < f_k(g), f_m(k_l^{-1}k_j g >= o(G)^{-1}. \sum_{h \in H} < f_k(h), f_m(k_l^{-1}k_j h) >$$

This is clearly zero if $l \neq j$ since then $k_l^{-1}k_j \notin H$, ie, the coset $k_l^{-1}k_j H$ is disjoint from H. On the other hand, when $l = j$, the above evaluates to

$$o(G)^{-1} \sum_{h \in H} < f_k(h), f_m(h) >= o(H)^{-1} \sum_{h \in H} < L(h)^{-1}\phi_k, L(h)^{-1}\phi_m >$$

$$=< \phi_k, \phi_m >= \delta_{k,m}$$

This completes the proof that $\mathcal{B}$ is indeed an onb for X. Now let C be a class in G. Let χ_U denote the character of U and χ_L the character of L. We define $\chi_U(C)$ to be the character χ_U of U evaluated at any element of C. Note that these are all equal. Then, we can write

$$C\bigcap H = \bigcup_{l=1}^{M} C_l$$

Where the $C_l's$ are disjoint classes in H. In the language of group algebras, we define

$$\mathcal{C} = \sum_{g\in C} g$$

Then,

$$g\mathcal{C}.g^{-1} = \mathcal{C}\forall g \in G$$

and

$$U(\mathcal{C}) = \sum_{g\in C} U(g)$$

And hence,

$$Tr(U(\mathcal{C})) = o(C).\chi_U(C)$$

on the one hand, while on the other,

$$Tr(U(\mathcal{C})) = \sum_{j,k} < U(k_j)f_k, U(\mathcal{C})U(k_j)f_k >$$

$$= \sum_{j,k} < f_k, U(k_j^{-1})U(\mathcal{C})U(k_j)f_k >$$

$$= \sum_{j,k} < f_k, U(\mathcal{C})f_k >$$

$$= (r/o(G)) \sum_{k,g\in C,h\in H} < f_k(h), f_k(g^{-1}h) >$$

$$= (rc^2/o(G)) \sum_{k,g\in C\cap H,h\in H} < L(h)^{-1}\phi_k, L(h^{-1}g)\phi_k >$$

$$= r. \sum_{k,g\in C_l,l} < \phi_k, L(g)\phi_k >= r. \sum_{g\in C_l,l} \chi_L(g) = r. \sum_{l=1}^{M} o(C_l)\chi_L(C_l)$$

Thus,

$$\chi_U(C) = (r/o(C)) \sum_{l=1}^{M} o(C_l)\chi_L(c_l).$$

Note that $r = o(G)/o(H) = c^2$ and M is the number of $H - classes$ in $C \cap H$ while $C_l, l = 1, 2, \ldots, M$ is the enumeration of these classes.

[3] The Frobenius reciprocity theorem. Let χ_1 be a character of G and $\tilde{\chi}_2$ a character of H. Let χ_2 denote the the character of G induced by $\tilde{\chi}_2$. Then we have by the Frobenius reciprocity theorem,

$$<\chi_1, \chi_2>= o(G)^{-1} \sum_{g \in G} \bar{\chi}_1(g)\chi_2(g)$$

$$= o(G)^{-1} \sum_{g \in C \in C(G), C_l \subset C} \bar{\chi}_1(g)(o(G)/o(H)o(C))o(C_l)\tilde{\chi}_2(C_l)$$

Where $C_l \subset C$ means that C_l ranges over all the H-classes in $C \bigcap H$ and $C(G)$ denotes the set of all the classes in G. We clearly have

$$\sum_{g \in C} \bar{\chi}_1(g) = o(C)\chi_1(C)$$

Thus, the above becomes

$$<\chi_1, \chi_2>=$$

$$o(H)^{-1} \sum_{C_l \subset C \in C(G)} o(C_l)\bar{\chi}_1(C)\tilde{\chi}_2(C_l)$$

$$= o(H)^{-1} \sum_{C_l \subset C \in C(G)} o(C_l)\bar{\chi}_1(C_l).\tilde{\chi}_2(C_l)$$

$$= o(H)^{-1} \sum_{C_l \in C(H)} o(C_l)\bar{\chi}_1(C_l).]\tilde{\chi}_2(C_l)$$

$$= o(H)^{-1}. \sum_{h \in H} \bar{\chi}_1(h)\tilde{\chi}_2(h) =< Res_H^G \chi_1, \tilde{\chi}_2 >$$

Where $Res_H^G \chi_1$ denotes the restriction of the function χ_1 on G to H. In particular, suppose χ_1 and $\tilde{\chi}_2$ are respectively irreducible representations of G and H. Then, the above formula can be expressed as

$$m_G(\chi_1, Ind_H^G \tilde{\chi}_2) = m_H(\tilde{\chi}_2, Res_H^G \chi_1)$$

where $m_G(\chi_1, \chi)$ is the multiplicity of the irreducible character χ_1 in the expansion of a character χ of G and $m_H(\tilde{\chi}_2, \tilde{\chi})$ is the multiplicity of the irreducible character $\tilde{\chi}_2$ in the expansion of a character $\tilde{\chi}$ of H. This the famous Frobenius reciprocity theorem.

[12] With reference to the previous part, in orthogonal curvilinear coordinates, q_1, q_2 for the plane, the electromagnetic field within a cavity resonator can be expressed as

$$E_z(t, q_1, q_2, z) = \sum_{n,p \geq 1} u_n(q_1, q_2).(2/d)^{1/2}.cos(\pi pz/d).Re(c(n,p)exp(-j\omega^E(n,p)t))$$

$$H_z(t, q_1, q_2, z) = \sum_{n,p \geq 1} v_n(q_1, q_2).(2/d)^{1/2}.sin(\pi pz/d).Re(d(n,p)exp(-j\omega^H(n,p)t))$$

$$\mathbf{E}_\perp(t,q_1,q_2,z)=\sum_{n,p} h_n^{-2}.(-\pi p/d).\nabla_\perp u_n(q_1,q_2).(2/d)^{1/2}.sin(p\pi z/d).Re(c(n,p).exp(-j\omega^E(n,p)t))$$

$$-\sum_{n,p} k_n^{-2}.(\nabla_\perp v_n)(q_1,q_2)\times\hat{z}).(2/d)^{1/2}.sin(p\pi z/d).Re(j\mu\omega^H(n,p)d(n,p).exp(-j\omega^H(n,p)t))$$

$$\mathbf{H}_\perp(t,q_1,q_2,z)=\sum_{n,p} k_n^{-2}.(\pi p/d).\nabla_\perp v_n(q_1,q_2).(2/d)^{1/2}.cos(p\pi z/d).Re(d(n,p).exp(-j\omega^H(n,p)t))$$

$$+\sum_{n,p} h_n^{-2}.(\nabla_\perp u_n)(q_1,q_2)\times\hat{z}).(2/d)^{1/2}.cos(p\pi z/d).Re(j\epsilon\omega^E(n,p)c(n,p).exp(-j\omega^E(n,p)t))$$

(see Exercise 2, in step 7). Show that the energy density in this confined electromagnetic field, assuming that u_n, v_n are normalized, can be expressed in the form

$$U=(\epsilon/2)\int|\mathbf{E}|^2dS(q_1,q_2)dz+(\mu/2)\int|\mathbf{H}|^2dS(q_1,q_2)dz$$

$$=(1/2)\sum_{n,p}[(\epsilon+\epsilon(\pi p/dh_n)^2)c(n,p,t)^2$$

$$+(\mu/h_n^2)(\tilde{c}(n,p,t)^2)+(\mu+\mu(\pi p/dk_n)^2)d(n,p,t)^2+(\epsilon/k_n^2)\tilde{d}(n,p,t)^2]$$

where

$$c(n,p,t)=Re(c(n,p)exp(-j\omega^E(n,p)t)),$$
$$\tilde{c}(n,p,t)=Re(j\epsilon\omega^E(n,p)c(n,p)exp(-j\omega^E(n,p)t)$$
$$d(n,p,t)=Re(d(n,p)exp(-j\omega^H(n,p)t)),$$
$$\tilde{d}(n,p,t)=Re(j\mu\omega^H(n,p)d(n,p)exp(-j\omega^H(n,p)t)$$

Show that U is time independent, ie, the total field energy in the cavity is conserved.

hint: Write

$$c(n,p)=c_R(n,p)+jc_I(n,p)$$

where c_R, c_I are real. Then,

$$(\epsilon+\epsilon(\pi p/dh_n))c(n,p,t)^2+(\mu/h_n^2)\tilde{c}(n,p,t)^2$$

$$(\epsilon+\epsilon(\pi p/dh_n)^2)(c_R(n,p)cos(\omega^E(n,p)t)+c_I(n,p)sin(\omega^E(n,p)t))^2$$
$$+(\mu/h_n^2)(\epsilon\omega^E(n,p))^2(c_R(n,p)sin(\omega^E(n,p)t)-c_I(n,p)cos(\omega^E(n,p)t))^2$$

Now observe that

$$\epsilon+\epsilon(\pi p/dh_n)^2=$$
$$(h_n^2+\pi^2p^2/d^2)\epsilon/h_n^2=\omega^E(n,p)^2\epsilon^2\mu/h_n^2$$

and further,

$$(\mu/h_n^2)(\epsilon\omega^E(n,p))^2=\omega^E(n,p)^2\epsilon^2\mu/h_n^2$$

and hence deduce that

$$(\epsilon+\epsilon(\pi p/dh_n)^2)c(n,p,t)^2+(\mu/h_n^2)\tilde{c}(n,p,t)^2=$$

$$(\omega^E(n,p)^2\epsilon^2\mu/h_n^2)(c_R(n,p)^2+c_I(n,p)^2)$$

Likewise, show that

$$(\mu+\mu(\pi p/dk_n)^2)d(n,p,t)^2+(\epsilon/k_n^2)\tilde{d}(n,p,t)^2$$

$$=(\omega^H(n,p)^2\mu^2\epsilon/k_n^2)(d_R(n,p)^2+d_I(n,p)^2)$$

Deduce that the energy of the field is given by

$$U=(1/2)\sum_{n,p}(\omega^E(n,p)a^E(n,p)^*a^E(n,p)+\omega^H(n,p)a^H(n,p)^*a^H(n,p))$$

where

$$a^E(n,p)=(\sqrt{\omega^E(n,p)}\epsilon\sqrt{\mu}/h_n)d(n,p)$$

$$a^H(n,p)=(\sqrt{\omega^H(n,p)}\mu\sqrt{\epsilon}/k_n)c(n,p)$$

(Recall that $c(n,p)=c_R(n,p)+jc_I(n,p), d(n,p)=d_R(n,p)+jd_I(n,p)$).

Explain how you would quantize this electromagnetic field based on the introduction of Bosonic creation and annihilation operators by interpreting the above field energy as the Hamiltonian of an infinite sequence of independent harmonic oscillators.

[13] Speech → MRI conversion using artificial neural networks. This problem outlines a procedure for predicting the MRI image data of a patient's brain from his slurred speech data using a combination of a feed-forward neural network and the extended Kalman filter. We assume that there is a definite relationship between the speech signal of a patient and his dynamically varying MRI image field. It should be noted that the speech data is a low dimensional signal, say of 100 time samples while the MRI is a much higher dimensional signal, again of say 100 time samples but each sample is a vector of a very large size. So we are predicting a very high dimensional data from a lower dimensional data and this enables us to avoid the use of expensive equipment for the purpose. Let $s(t)$ denote the speech signal of the patient and $\mathbf{f}(t)$ the MRI image data of the same patient transformed from a matrix image field to a vector via the Vec operation. It is assumed that $\mathbf{f}(t)=[f_1(t),...,f_L(t)]^T$ can be expressed as some function of $\mathbf{s}(t)=[s(t),s(t-1),...,s(t-L)]^T$. The neural network is assumed to have K layers with each layer having N nodes. Let $W(k,l,m), 1<k\leq K, 1\leq l,m\leq N$ denote the weights connecting the $(k-1)^{th}$ layer to the kth layer. The input speech signal $\mathbf{s}(t)$ is applied at the zeroth layer. Thus, writing the weight matrices as

$$\mathbf{W}(k)=((W(k,l,m)))_{1\leq l,m\leq N}\in\mathbb{R}^{N\times N}$$

it follows that the signal vector at the kth layer is given by

$$\mathbf{x}_k(t)=\sigma(\mathbf{W}(k)\mathbf{x}_{k-1}(t)), k=1,2,...,L$$

where

$$\mathbf{x}_0(t) = \mathbf{s}(t)$$

and the output vector of the network

$$\mathbf{y}(t) = \mathbf{x}_L(t)$$

is matched to the MRI process $\mathbf{f}(t)$ at each time. Here σ is the sigmoidal function that acts on each components of its argument, ie, if we write

$$\mathbf{W}(k)\mathbf{x}_{k-1}(t) = \mathbf{z}(t) = [z_1(t), ..., z_L(t)]^T,$$

then

$$\sigma(\mathbf{W}(k)\mathbf{x}_{k-1}(t)) = [\sigma(z_1(t)), ..., \sigma(z_L(t))]^T$$

The measurement data at time t is the neural network output $\mathbf{y}(t)$ plus noise/error which we assume to be equal to the true MRI process $\mathbf{f}(t)$. To estimate the weights of the neural network, we assume a weight dynamics

$$Vec(\mathbf{W}(k)(n+1)) = Vec(\mathbf{W}(k))(n) + \epsilon_W(k)(n)$$

where ϵ_W is a noise/error process. We write the measurement model as

$$\mathbf{f}(t) = \mathbf{y}(t) + v(t) =$$

$$= \sigma(\mathbf{W}(K)(t)...\sigma(\mathbf{W}(2)(t)\sigma(\mathbf{W}(1)(t)\mathbf{s}(t)))) + v(t)$$
$$= \mathbf{h}(\mathbf{W}(t), \mathbf{s}(t)) + v(t)$$

where

$$\mathbf{W}(t) = [Vec(\mathbf{W}(1))(t)^T, ..., Vec(\mathbf{W}(K)(t)^T]^T$$

is the vector of all the weights at time t. The EKF is driven by the output MRI signal $\mathbf{f}(t)$ and noting that the weight dynamics can be expressed in the form

$$\mathbf{W}(t+1) = \mathbf{W}(t) + \epsilon_W(t)$$

it follows that the EKF can be cast in the form

$$\hat{\mathbf{W}}(t+1|t) = \hat{\mathbf{W}}(t|t),$$
$$\hat{\mathbf{W}}(t+1|t+1) = \hat{\mathbf{W}}(t+1|t) + \mathbf{K}(t+1)(\mathbf{f}(t+1) - \mathbf{h}(\hat{\mathbf{W}}(t+1|t), \mathbf{s}(t+1))$$
$$\mathbf{K}(t+1) = \mathbf{P}(t+1|t)\mathbf{H}(t+1)^T(\mathbf{H}(t+1)^T\mathbf{P}(t+1|t)\mathbf{H}(t+1) + \mathbf{P}_v)^{-1}$$
$$\mathbf{H}(t+1) = \frac{\partial \mathbf{h}(\hat{\mathbf{W}}(t+1|t), \mathbf{s}(t+1))}{\partial \mathbf{W}}$$
$$\mathbf{P}(t+1|t) = \mathbf{P}(t|t) + P_\epsilon,$$
$$\mathbf{P}(t+1|t+1) = \mathbf{I} - \mathbf{K}(t+1)\mathbf{H}(t+1))\mathbf{P}(t+1|t)(\mathbf{I} - \mathbf{K}(t+1)\mathbf{H}(t+1))^T$$
$$+\mathbf{K}(t+1)\mathbf{P}_\epsilon\mathbf{K}(t+1)^T$$

Derive these equations from first principles and implement this on MATLAB. For MATLAB implementation, you can use a two layered neural network. The problem of computing the Jacobian $\mathbf{H}$ of $\mathbf{h}(\mathbf{W}, \mathbf{s})$ involves using back-propagation which is an elementary application of the chain rule of differential calculus, ie, writing

$$\mathbf{x}_k = \sigma(\mathbf{W}(k)\mathbf{x}_{k-1}), k = 1, 2, ..., L$$

we have

$$\partial \mathbf{h}(\mathbf{W}, \mathbf{s})/\partial Vec(\mathbf{W}(k)) = \partial \mathbf{y}/\partial Vec(\mathbf{W}(k)) =$$

$$\frac{\partial \mathbf{y}}{\partial \mathbf{x}_{L-1}} \cdot \frac{\partial \mathbf{x}_{L-1}}{\partial \mathbf{x}_{L-2}} \cdots \frac{\partial \mathbf{x}_{k+1}}{\partial \mathbf{x}_k} \cdot \frac{\partial \mathbf{x}_k}{\partial Vec\mathbf{W}(k)}$$

Write down all the terms involved in this expression explicitly for this model.

Approximation of multivariate polynomials using a neural network: In order to characterize the performance of a neural network in approximating a given plant function, we require to calculate the mean square approximation error involved in approximating polynomial functions using the network. The sigmoidal functions used in the network must be approximated again by polynomials based on truncating their Taylor series upto a given degree N and then the minimum mean square estimation error evaluated based on such an approximation. Consider first a two layer neural network with each layer having two nodes. The output vector is then

$$\mathbf{y} = \sigma(\mathbf{W}_2\sigma(\mathbf{W}_1\mathbf{x}))$$

with $\mathbf{x} = [x_1, x_2]^T$ as the input and $\mathbf{W}_k, k - 1, 2$ are 2×2 matrices. Therefore in component form,

$$y_1 = [\sigma(W_2(11)z_1 + W_2(12)z_2), \sigma(W_2(21)z_1 + W_2(22)z_2)]^T,$$

where

$$z_1 = \sigma(W_1(11)x_1 + W_1(12)x_2), z_2 = \sigma(W_1(21)x_1 + W_1(22)x_2)$$

The aim is to compute the minimum mean square approximation error

$$min_{\mathbf{W}_1,\mathbf{W}_2} \int_D ((p_1(x_1, x_2) - y_1)^2 + (p_2(x_1, x_2) - y_2)^2) dx_1 dx_2$$

over a domain $D \subset \mathbb{R}^2$ for a given bivariate polynomials p_1, p_2. For example, we can take $D = [a, b] \times [c, d]$. This problem is the same as minimizing the mean square error between the random variables $\mathbf{p}(\mathbf{x})$ and $\mathbf{y} = \mathbf{y}(\mathbf{x})$ when $\mathbf{x}$ is a uniformly distributed random vector on D. More generally, we can talk about minimizing this mean square error when $\mathbf{x}$ has any given probability distribution $F(\mathbf{x})$ in $\mathbb{R}^2$. Writing

$$\sigma(z) \approx \sum_{k=0}^{N} c(k) z^k$$

we have that

$$\sigma(a_1u_1 + a_2u_2) = \sum_{k=0}^{N} c(k)(a_1u_1 + a_2u_2)^k$$

$$= \sum_{0\le m\le k\le N} c(k)\binom{k}{m} a_1^m a_2^{k-m} u_1^m u_2^{k-m}$$

Thus,

$$z_1 = \sum_{k,m} c(k)\binom{k}{m} W_1(11)^m W_1(12)^{k-m} x_1^m x_2^{k-m}$$

$$z_2 = \sum_{k,m} c(k)\binom{k}{m} W_1(21)^m W_1(22)^{k-m} x_1^k x_2^{k-m}$$

$$y_1 = \sum_{k,m} c(k)\binom{k}{m} W_2(11)^m W_2(12)^{k-m} z_1^m z_2^{k-m}$$

$$= \sum c(k)\binom{k}{m} W_2(11)^m W_2(12)^{k-m} c(k_1)...c(k_m)\binom{k_1}{n_1}...\binom{k_m}{n_m}$$

$$\times W_1(11)^{n_1+...+n_m} W_1(12)^{k_1+...+k_n-n_1-...-n_n} x_1^{n_1+...+n_m} x_2^{k_1+...+k_n-m_1-...-m_n}$$

For fixed x_1, x_2, this is a polynomial in $(W_1(11), W_1(12))$ of degree Nn. Likewise, the formula for y_2 is given by

$$y_2 = \sum_{k,m} c(k)\binom{k}{m} W_2(21)^m W_2(22)^{k-m} z_1^m z_2^{k-m}$$

$$= \sum c(k)\binom{k}{m} W_2(21)^m W_2(22)^{k-m} c(k_1)...c(k_m)\binom{k_1}{n_1}...\binom{k_m}{n_m}$$

$$\times W_1(11)^{n_1+...+n_m} W_1(12)^{k_1+...+k_n-n_1-...-n_n} x_1^{n_1+...+n_m} x_2^{k_1+...+k_n-m_1-...-m_n}$$

Choosing $D = [0,1] \times [0,1]$ without loss of generality (since we can in the case when $D = [a,b] \times [a,b]$ replace the sigmoidal function σ by a scaled and translated version of it thereby reducing the problem to $[0,1]\times[0,1]$. To calculate the mean squared approximation error between $(y_1(x_1,x_2), y_2(x_1,x_2))$ and a bivariate polynomial pair $(p_1(x_1,x_2), p_2(x_1,x_2))$, we need to evaluate the following integrals:

$$\int_0^1\int_0^1 y_1^2 dx_1 dx_2, \int_0^1\int_0^1 y_2^2 dx_1 dx_2,$$

$$\int_0^1\int_0^1 y_1 x_1^r x_2^s dx_1 dx_2, \int_0^1\int_0^1 y_2 x_1^r x_2^s dx_1 dx_2$$

We have

$$\int_0^1\int_0^1 y_1 x_1^r x_2^s dx_1 dx_2 =$$

Acknowledgement for problem [13]: I've borrowed this problem from my informal discussions with Prof.Vijayant Agarwal, Vijay Upreti and Gugloth Sagar.

[14] This problem outlines the steps involved in deriving the EKF and UKF in discrete time.

step 1: Let the state vector $\mathbf{x}(t) \in \mathbb{R}^N$ satisfy the stochastic difference equation

$$\mathbf{x}(t+1) = \mathbf{f}(t, \mathbf{x}(t)) + \mathbf{g}(t, \mathbf{x}(t))\mathbf{w}(t+1), t = 0, 1, 2, ... - - - (1)$$

where

$$\mathbf{f} : \mathbb{R}_+ \times \mathbb{R}^N \to \mathbb{R}^N, \mathbf{g} : \mathbb{R}_+ \times \mathbb{R}^N \to \mathbb{R}^{N \times p}$$

and $\mathbf{w}(t), t = 1, 2, ...$ white noise with zero mean. Its autocorrelation is given by

$$\mathbb{E}(\mathbf{w}(t)\mathbf{w}(s)^T) = \mathbf{Q}(t)\delta[t-s]$$

The measurement model is

$$\mathbf{z}(t) = \mathbf{h}(t, \mathbf{x}(t)) + \mathbf{v}(t), t = 0, 1, 2, ... - - - (2)$$

where $\mathbf{v}(.)$ is also zero mean white and independent of $\mathbf{w}(.)$:

$$\mathbb{E}(\mathbf{v}(t)\mathbf{v}(s)^T) = \mathbf{R}(t)\delta[t-s],$$

$$\mathbb{E}(\mathbf{w}(t)\mathbf{v}(s)^T) = \mathbf{0}$$

(1) is called the state/process model and (2) the measurement model. $\mathbf{w}(.)$ is called process noise and $\mathbf{v}(.)$ called measurement noise. Note that when we say white noise, we mean that its samples are statistically independent, not just uncorrelated. If the noise is Gaussian, then uncorrelatedness implies independence but for non-Gaussian noises, independence is a stronger condition than uncorrelatedness. Let

$$\mathbf{Z}_t = \{\mathbf{z}(s) : 0 \leq s \leq t\}$$

the measurement data collected upto time t. We shall be assuming that $\mathbf{x}(0)$ is independent of $\{\mathbf{w}(t), \mathbf{v}(t) : t \geq 1\}$

Remark: At time t, we have available with us $\hat{\mathbf{x}}(t|t) = \mathbb{E}(\mathbf{x}(t)|\mathbf{Z}_t)$ and $\mathbf{P}(t|t) = Cov(\mathbf{e}(t|t)|\mathbf{Z}_t) = Cov(\mathbf{x}(t)|\mathbf{Z}_t)$ where $\mathbf{e}(t|t) = \mathbf{x}(t) - \hat{\mathbf{x}}(t|t)$. The iteration process involves computing $\hat{\mathbf{x}}(t+1|t+1)$ and $\mathbf{P}(t+1|t+1)$. This computation progresses in two stages, first compute $\hat{\mathbf{x}}(t+1|t) = \mathbb{E}(\mathbf{x}(t+1)|\mathbf{Z}_t)$, $\mathbf{e}(t+1|t) = \mathbf{x}(t+1) - \hat{\mathbf{x}}(t+1|t)$ and $\mathbf{P}(t+1|t) = Cov(\mathbf{e}(t+1|t)|\mathbf{Z}_t) = Cov(\mathbf{x}(t+1)|\mathbf{Z}_t)$ based on only the state dynamics, and then update these to $\hat{\mathbf{x}}(t+1|t+1)$ and $\mathbf{P}(t+1|t+1)$ based on the measurement $\mathbf{z}(t+1)$.

step (2a):Computation of $\hat{\mathbf{x}}(t+1|t)$.

Then from the state model, we have that $\mathbf{x}(t)$ is a function of $\mathbf{x}(0), \mathbf{w}(1), ..., \mathbf{w}(t-1)$ and $\mathbf{z}(t)$ is hence a function of $\mathbf{x}(0), \mathbf{w}(1), ..., \mathbf{w}(t-1), \mathbf{v}(t)$. Thus, $\mathbf{Z}_t$ is a function of $\mathbf{x}(0), \mathbf{w}(1), ..., \mathbf{w}(t-1), \mathbf{v}(0), ...\mathbf{v}(t)$. It then follows from the assumptions made that $\mathbf{w}(t+1)$ is independent of $(\mathbf{Z}_t, \mathbf{x}(t))$. Hence, taking conditional expectations on both sides of of (1), and using the independence of $\mathbf{w}(t+1)$ and $(\mathbf{Z}_t, \mathbf{x}(t))$, we get

$$\hat{\mathbf{x}}(t+1|t) = \mathbb{E}(\mathbf{x}(t+1)|\mathbf{Z}_t) =$$

$$\mathbb{E}(\mathbf{f}(t, \mathbf{x}(t))|\mathbf{Z}_t) + \mathbb{E}(\mathbf{g}(t, \mathbf{x}(t))\mathbf{w}(t+1)|\mathbf{Z}_t)$$

where

$$\mathbb{E}[(\mathbf{g}(t, \mathbf{x}(t))\mathbf{w}(t+1)|\mathbf{Z}_t]$$

$$= \mathbb{E}[\mathbb{E}[(\mathbf{g}(t, \mathbf{x}(t))\mathbf{w}(t+1)|\mathbf{Z}_t, \mathbf{x}(t)]|\mathbf{Z}_t]$$

$$= \mathbb{E}[\mathbf{g}(t, \mathbf{x}(t))\mathbb{E}[\mathbf{w}(t+1)|\mathbf{Z}_t, \mathbf{x}(t)]|Z_t] = \mathbf{0}$$

since

$$\mathbb{E}[\mathbf{w}(t+1)|\mathbf{Z}_t, \mathbf{x}(t)] = \mathbb{E}\mathbf{w}(t+1) = \mathbf{0}$$

Remark: if $\mathbf{x}, \mathbf{y}, \mathbf{z}$ are random vectors, then

$$\mathbb{E}(\mathbb{E}(\mathbf{x}|\mathbf{y}, \mathbf{z})|\mathbf{z}) = \mathbb{E}(\mathbf{x}|\mathbf{z})$$

and if $\mathbf{x}, \mathbf{y}$ are independent, then

$$\mathbb{E}(\mathbf{x}|\mathbf{y}) = \mathbb{E}(\mathbf{x})$$

Prove these statements by assuming joint probability densities. Thus, we get

$$\hat{\mathbf{x}}(t+1|t) = \mathbb{E}(\mathbf{f}(t, \mathbf{x}(t))|\mathbf{Z}_t)$$

Now if $\mathbf{f}(t, \mathbf{x})$ is affine linear in $\mathbf{x}$), ie, of the form $\mathbf{u}(t) + \mathbf{A}(t)\mathbf{x}$, then it is immediate that

$$\mathbb{E}((\mathbf{f}(t, \mathbf{x}(t))|\mathbf{Z}_t) = \mathbf{f}(t, \mathbb{E}(\mathbf{x}(t)|\mathbf{Z}_t))$$

$$= \mathbf{f}(t, \hat{\mathbf{x}}(t|t))$$

In the general case, however, we cannot make this assumption. If $\mathbf{f}(t, \mathbf{x})$ is analytic in $\mathbf{x}$, then we can Taylor expand it around $\hat{x}(t|t)$

$$\mathbf{f}(t, \mathbf{x}) = \mathbf{u}(t) + f_x(t, \hat{x}(t|t)) + \sum_{n\geq 1} \mathbf{A}_n(t)(\mathbf{x}) - \hat{x}(t|t))^{\otimes n}/n!$$

and then get

$$\mathbb{E}[\mathbf{f}(t, \mathbf{x}(t))|\mathbf{Z}_t] = \mathbf{u}(t) + \sum_{n\geq 2} \mathbf{A}_n(t)\mu_n(t|t)/n!$$

where

$$\mathbf{u}(t) = \mathbf{f}(t, \hat{x}(t|t))$$

and $\mu_n(t|t)$ is the conditional n^{th} order estimation error moment of $\mathbf{x}(t)$ given $\mathbf{Z}_t$, ie,

$$\mu_n(t) = \mathbb{E}((\mathbf{x}(t) - \hat{x}(t|t))^{\otimes n}|\mathbf{Z}_t)$$

However if we were to implement a filter based on such an approach, it will become an infinite dimensional filter, ie, at each time step, we have to update the conditional moments of all orders. The EKF is a finite dimensional approximation to such an infinite dimensional filter in which we neglect the conditional moments of all orders greater than two. Thus, we get an approximation

$$\mathbb{E}[\mathbf{f}(t,\mathbf{x}(t))|\mathbf{Z}_t] \approx f(t, \hat{\mathbf{x}}(t|t)) + (1/2)f_{,xx}(t, \hat{\mathbf{x}}(t|t)Vec(P(t|t))$$

We note that

$$\mu_2(t|t) = Vec(P(t|t))$$

Most authors also neglect the second term here and simply make the approximation

$$\mathbb{E}[\mathbf{f}(t,\mathbf{x}(t))|\mathbf{Z}_t] \approx \mathbf{f}(t, \hat{\mathbf{x}}(t|t))$$

The UKF on the other hand, gives a better approximation even than that obtained by truncating in the above way upto a given order of moments. It is based on evaluating the conditional expectation $\mathbb{E}(f(t,\mathbf{x}(t))|Z_t)$ using the law of large numbers. Specifically writing

$$\mathbf{e}(t|t) = \mathbf{x}(t) - \hat{\mathbf{x}}(t|t)$$

and defining

$$P(t|t) = Cov(\mathbf{e}(t|t)|\mathbf{Z}_t) = Cov(\mathbf{x}(t)|\mathbf{Z}_t)$$

we choose a sequence $\xi(m), m = 1, 2, ..., K$ of iid $N(\mathbf{0}, \mathbf{I}_N)$ random vectors independent of $\mathbf{Z}_t$ and note that by the law of large numbers,

$$(1/K)\sum_{m=1}^{K} \mathbf{f}(t, \hat{\mathbf{x}}(t|t) + \sqrt{\mathbf{P}(t|t)}\xi(m))$$

conditioned on $\mathbf{Z}_t$ converges to $\mathbb{E}(\mathbf{f}(t,\mathbf{x}(t))|\mathbf{Z}_t)$ as $K \to \infty$ provided that conditioned on $\mathbf{Z}_t$, $\mathbf{e}(t|t)$ has a normal distribution. Thus, the result of the UKF is

$$\hat{\mathbf{x}}(t+1|t) = (1/K)\sum_{m=1}^{K} \mathbf{f}(t, \hat{\mathbf{x}}(t|t) + \sqrt{\mathbf{P}(t|t)}\xi(m))$$

Note that this is also an approximation. However, if we assume that $\mathbf{e}(t|t)$ conditioned on $\mathbf{Z}_t$ has a probability distribution $F_{e,t|t}(\mathbf{e})$ and we choose $\zeta(m), m = 1, 2, ..., K$ conditioned on $\mathbf{Z}_t$ to be iid with this distribution $F_{e,t|t}(\mathbf{e})$, then by the law of large numbers, $(1/K)\sum_{m=1}^{K} \mathbf{f}(t, \hat{\mathbf{x}}(t) + \zeta(m))$ will converge as $K \to \infty$ conditioned on $\mathbf{Z}_t$ to $\mathbb{E}(\mathbf{f}(t,\mathbf{x}(t))|\mathbf{Z}_t)$. This defines the first stage of the EKF and the UKF.

step (2b): We now have to compute

$$\mathbf{P}(t+1|t) = cov(\mathbf{x}(t+1)|\mathbf{Z}_t) = Cov(\mathbf{x}(t+1)-\hat{\mathbf{x}}(t+1|t)|\mathbf{Z}_t) = Cov(\mathbf{e}(t+1|t)|\mathbf{Z}_t)$$

The EKF computes this using the following:

$$\mathbf{e}(t+1|t) = \mathbf{x}(t+1) - \mathbf{f}(t, \hat{\mathbf{x}}(t|t)) =$$

$$\mathbf{f}(t, \mathbf{x}(t)) + \mathbf{g}(t, \mathbf{x}(t))\mathbf{w}(t+1) - \mathbf{f}(t, \hat{\mathbf{x}}(t|t))$$

$$\approx \mathbf{f}_{,x}(t, \hat{\mathbf{x}}(t|t))\mathbf{e}(t|t) + \mathbf{g}(t, \hat{\mathbf{x}}(t|t)).\mathbf{w}(t+1)$$

and hence

$$\mathbf{P}(t+1|t) = \mathbf{f}_{,x}(t, \hat{\mathbf{x}}(t|t))\mathbf{P}(t|t)\mathbf{f}_{,x}(t, \hat{\mathbf{x}}(t|t))^T + \mathbf{g}(t, \hat{\mathbf{x}}(t|t))\mathbf{Q}(t).\mathbf{g}(t, \hat{\mathbf{x}}(t|t))^T$$

In the UKF, we are not allowed to make the approximation $\mathbf{f}(t, \hat{\mathbf{x}}(t|t))$ for $\mathbb{E}(\mathbf{f}(t, \mathbf{x}(t))|\mathbf{Z}_t)$. Instead we must use the independent realizations $\mathbf{f}(t, \hat{\mathbf{x}}(t|t) + \sqrt{\mathbf{P}(t|t)}\xi(m)), m = 1, 2, ..., K$ of $\mathbf{f}(t, \mathbf{x}(t)) = \mathbf{f}(t, \hat{\mathbf{x}}(t|t) + \mathbf{e}(t|t))$ conditioned on $\mathbf{Z}_t$. Thus, in the UKF, based on the large numbers, we compute

$$\mathbf{P}(t+1|t) =$$

$$(1/K)\sum_{m=1}^{K}(\mathbf{f}(t, \hat{\mathbf{x}}(t|t)+\sqrt{\mathbf{P}(t|t)}\xi(m))-\hat{\mathbf{x}}(t+1|t)).(\mathbf{f}(t, \hat{\mathbf{x}}(t|t)+\sqrt{\mathbf{P}(t|t)}\xi(m))-\hat{\mathbf{x}}(t+1|t))^T$$

step 3: The EKF computation of $\hat{\mathbf{x}}(t+1|t+1)$ and $\mathbf{P}(t+1|t+1)$.

$$\hat{\mathbf{x}}(t+1|t+1) = \mathbb{E}(\mathbf{x}(t+1)|\mathbf{Z}_{t+1}) = \mathbb{E}(\mathbf{x}(t+1)|\mathbf{Z}_t, \mathbf{z}(t+1))$$

In the EKF, we assume that the extra measurement modifies $\hat{\mathbf{x}}(t+1|t)$ by an additive term proportional to the output error at time $t+1$, ie, the difference between the true output measurement $\mathbf{z}(t+1)$ and its estimate $\hat{\mathbf{z}}(t+1|t) = \mathbb{E}(\mathbf{h}(t+1, \mathbf{x}(t+1))|\mathbf{Z}_t)$ based on $\mathbf{Z}_t$. Thus the EKF gives

$$\hat{\mathbf{x}}(t+1|t+1) = \hat{\mathbf{x}}(t+1|t) + \mathbf{K}(t+1)(\mathbf{z}(t+1) - \mathbf{h}(t, \hat{\mathbf{x}}(t+1|t)))$$

where $\mathbb{E}(\mathbf{h}(t+1, \mathbf{x}(t+1))|\mathbf{Z}_t)$ has been approximated by $\mathbf{h}(t+1, \mathbb{E}\mathbf{x}(t+1)|\mathbf{Z}_t)) = \mathbf{h}(t+1, \hat{\mathbf{x}}(t+1|t))$ Again the conditional expectation has been pushed inside the nonlinearity. This algorithm for updating the conditional expectation based on the newly arrived measurement is based on the fact that if at time t, the output estimation error is "positive", then we increase proportionally the state estimate while if it is negative, we decrease proportionally the state estimate.

The "Kalman gain" $\mathbf{K}(t+1)$ is computed so that

$$\mathbb{E}(\| \hat{\mathbf{x}}(t+1) - \hat{\mathbf{x}}(t+1|t+1) \|^2 |\mathbf{Z}_t) = \mathbb{E}[\| \mathbf{e}(t+1|t+1) \|^2 |\mathbf{Z}_t]$$

is a minimum. We note that

$$\mathbf{e}(t+1|t+1) = \mathbf{x}(t+1) - \hat{\mathbf{x}}(t+1|t+1) =$$

$$\begin{aligned}&\mathbf{x}(t+1)-\hat{\mathbf{x}}(t+1|t)-\mathbf{K}(t+1)(\mathbf{z}(t+1)-\mathbf{h}(t,\hat{\mathbf{x}}(t+1|t)))\\&=\mathbf{e}(t+1|t)-\mathbf{K}(t+1)(\mathbf{h}(t+1,\mathbf{x}(t+1))-\mathbf{h}(t,\hat{\mathbf{x}}(t+1|t))+\mathbf{v}(t+1))\\&\approx(\mathbf{I}-\mathbf{K}(t+1)\mathbf{H}(t+1))\mathbf{e}(t+1|t)+\mathbf{K}(t+1)\mathbf{v}(t+1)\end{aligned}$$

where

$$\mathbf{H}(t+1)=\frac{\partial\mathbf{h}(t+1,\hat{\mathbf{x}}(t+1|t))}{\partial\mathbf{x}}$$

and hence

$$\begin{aligned}&E[\|\ \mathbf{e}(t+1|t+1)\ \|^2\ |\mathbf{Z}_t]=\\&=Tr[(\mathbf{I}-\mathbf{K}(t+1)\mathbf{H}(t+1))\mathbf{P}(t+1|t)(\mathbf{I}-\mathbf{K}(t+1)\mathbf{H}(t+1))^T\\&+\mathbf{K}(t+1)\mathbf{R}(t+1)\mathbf{K}(t+1)^T]\end{aligned}$$

Minimizing this w.r.t $\mathbf{K}(t+1)$ using the variational calculus for functions of matrices gives us the optimum Kalman gain as

$$\mathbf{K}(t+1)=\mathbf{P}(t+1|t)\mathbf{H}(t+1)^T(\mathbf{H}(t+1)\mathbf{P}(t+1|t)\mathbf{H}(t+1)^T+\mathbf{R}(t+1))^{-1}$$

The optimum value of $\mathbf{P}(t+1|t+1)$ is then obtained by subsituting this value of the optimal Kalman gain into the expression

$$\begin{aligned}&E[\mathbf{e}(t+1|t+1)\mathbf{e}(t+1|t+1)^T|\mathbf{Z}_t]=\\&=[(\mathbf{I}-\mathbf{K}(t+1)\mathbf{H}(t+1))\mathbf{P}(t+1|t)(\mathbf{I}-\mathbf{K}(t+1)\mathbf{H}(t+1))^T\\&+\mathbf{K}(t+1)\mathbf{R}(t+1)\mathbf{K}(t+1)^T]\end{aligned}$$

and assuming that $\mathbf{P}(t+1|t+1)$ is a function of only $\mathbf{Z}_t$ and not of $\mathbf{Z}_{t+1}=(\mathbf{Z}_t,\mathbf{z}(t+1))$. The above expression then yields

$$\begin{aligned}\mathbf{P}(t+1|t+1)&=[(\mathbf{I}-\mathbf{K}(t+1)\mathbf{H}(t+1))\mathbf{P}(t+1|t)(\mathbf{I}-\mathbf{K}(t+1)\mathbf{H}(t+1))^T\\&+\mathbf{K}(t+1)\mathbf{R}(t+1)\mathbf{K}(t+1)^T]\\&=(\mathbf{I}-\mathbf{K}(t+1)\mathbf{H}(t+1))\mathbf{P}(t+1|t)\\&=\mathbf{P}(t+1|t)-\mathbf{P}(t+1|t)\mathbf{H}(t+1)^T(\mathbf{H}(t+1)\mathbf{P}(t+1|t)\mathbf{H}(t+1)^T\\&+\mathbf{R}(t+1))^{-1}\mathbf{H}(t+1)\mathbf{P}(t+1|t)\end{aligned}$$

Exercise: Rewrite the expression for $\mathbf{P}(t+1|t+1)$ using the matrix inversion lemma.

Remark: Note that the above expression for $\mathbf{P}(t+1|t+1)$ shows that

$$\mathbf{P}(t+1|t+1)\leq\mathbf{P}(t+1|t)$$

which in particular implies that

$$Tr(\mathbf{P}(t+1|t+1))\leq Tr(\mathbf{P}(t+1|t))$$

This means that if we base our state estimate on an extra data point, the estimation error variance reduces. This is natural to expect.

step 4: The UKF calculation of $\hat{\mathbf{x}}(t+1|t+1)$ and $\mathbf{P}(t+1|t+1)$.
Here, we start with the expression

$$\hat{\mathbf{x}}(t+1|t+1) = \mathbb{E}[\mathbf{x}(t+1)|\mathbf{Z}_t, \mathbf{z}(t+1)]$$

and assume that given $\mathbf{Z}_t$, $(\mathbf{x}(t+1), \mathbf{z}(t+1))$ is jointly Gaussian. This is justified since $\mathbf{x}(t+1) = \hat{\mathbf{x}}(t+1|t) + \mathbf{e}(t+1|t)$ and $\mathbf{z}(t+1) = \hat{\mathbf{z}}(t+1|t) + \mathbf{e}_z(t+1|t)$ which means that the assumption amounts to saying that given $\mathbf{Z}_t$, $(\mathbf{e}(t+1|t), \mathbf{e}_z(t+1|t))$ are jointly Gaussian errors. Next, we use the fact that if $\mathbf{U}, \mathbf{V}$ are jointly Gaussian vectors, then

$$\mathbb{E}(\mathbf{X}|\mathbf{Y}) = \mu_X + \Sigma_{XY}\Sigma_{YY}^{-1}(\mathbf{Y} - \mu_Y),$$

$$Cov(\mathbf{X}|\mathbf{Y}) = \Sigma_{XX} - \Sigma_{XY}\Sigma_{YY}^{-1}\Sigma_{XY}^T$$

where

$$\Sigma_{XX} = Cov(\mathbf{X}), \Sigma_{YY} = Cov(\mathbf{Y}), \Sigma_{XY} = Cov(\mathbf{X}, \mathbf{Y}),$$

$$\mu_X = \mathbb{E}(\mathbf{X}), \mu_Y = \mathbb{E}(\mathbf{Y})$$

Based on these assumptions and formulae, we have the approximate formulae

$$\hat{\mathbf{x}}(t+1|t+1) = \mathbb{E}(\mathbf{x}(t+1)|\mathbf{Z}_t, \mathbf{z}(t+1)) =$$

$$\hat{\mathbf{x}}(t+1|t) - Cov(\mathbf{x}(t+1), \mathbf{z}(t+1)|\mathbf{Z}_t).Cov(\mathbf{z}(t+1)|\mathbf{Z}_t)^{-1}(\mathbf{z}(t+1) - \hat{\mathbf{z}}(t+1|t))$$

where

$$Cov(\mathbf{x}(t+1), \mathbf{z}(t+1)|\mathbf{Z}_t) = Cov(\mathbf{x}(t+1), \mathbf{h}(t+1, \mathbf{x}(t+1))|\mathbf{Z}_t)$$

$$= Cov(\mathbf{e}(t+1|t), \mathbf{h}(t+1, \hat{\mathbf{x}}(t+1|t) + \mathbf{e}(t+1|t))|\mathbf{Z}_t)$$

$$= (1/K)\sum_{m=1}^{K} \sqrt{\mathbf{P}(t+1|t)}\eta(m)(\mathbf{h}(t+1, \hat{\mathbf{x}}(t+1|t) + \sqrt{\mathbf{P}(t+1|t)}\eta(m)) - \hat{\mathbf{z}}(t+1|t))^T$$

where

$$\hat{\mathbf{z}}(t+1|t) = (1/K)\sum_{m=1}^{K} \mathbf{h}(t+1, \hat{\mathbf{x}}(t+1|t) + \sqrt{\mathbf{P}(t+1|t)}\eta(m))$$

where $\eta(m), m = 1, 2, ..., K$ are again iid standard normal random vectors. Moreover,

$$Cov(\mathbf{z}(t+1)|\mathbf{Z}_t) = Cov(\mathbf{h}(t+1, \mathbf{x}(t+1))|\mathbf{Z}_t) + \mathbf{R}(t+1)$$

$$= Cov(\mathbf{h}(t+1, \hat{\mathbf{x}}(t+1|t) + \mathbf{e}(t+1|t))|\mathbf{Z}_t) + \mathbf{R}(t+1)$$

$$= (1/K)\sum_{m=1}^{K}(\mathbf{h}(t+1, \hat{\mathbf{x}}(t+1|t) + \sqrt{\mathbf{P}(t+1|t)}\eta(m))$$

$$-\hat{\mathbf{z}}(t+1|t)).(\mathbf{h}(t+1, \hat{\mathbf{x}}(t+1|t) + \sqrt{\mathbf{P}(t+1|t)}\eta(m)) - \hat{\mathbf{z}}(t+1|t))^T$$

$$+\mathbf{R}(t+1)$$

and

$$\mathbf{P}(t+1|t+1) = Cov(\mathbf{x}(t+1)|\mathbf{Z}_t, \mathbf{z}(t+1))$$

$$= Cov(\mathbf{x}(t+1)|\mathbf{Z}_t) - Cov(\mathbf{x}(t+1),$$

$$\mathbf{z}(t+1)|\mathbf{Z}_t).Cov(\mathbf{z}(t+1)|\mathbf{Z}_t)^{-1}.Cov(\mathbf{x}(t+1), \mathbf{z}(t+1)|\mathbf{Z}_t)^T$$

where all the terms except the first on the rhs have been computed above. The first term is

$$Cov(\mathbf{x}(t+1)|\mathbf{Z}_t) = \mathbf{P}(t+1|t)$$

This completes the description of the EKF and the UKF.

step 5: Performance analysis of the UKF based on the large deviation principle.

[15] The Belavkin filter and how it improves upon the classical Kushner-Kallianpur filter. The unitary evolution in system $\otimes$ bath space is given by the Hudson-Parthasarathy noisy Schrodinger equation

$$dU(t) = [-(iH+P)dt - LdA + L^*dA^*]U(t), P = LL^*/2$$

For any system space observable X, we define the system state at time t to be

$$X(t) = j_t(X) = U(t)^*XU(t)$$

and by quantum Ito's formula, obtain

$$dj_t(X) = j_t(\theta_0(X))dt + j_t(\theta_1(X))dA(t) + j_t(\theta_2(X))dA(t)^*$$

where

$$\theta_0(X) = i[H,X] - XP - PX + LXL^* = i[H,X] - (1/2)(LL^*X + XLL^* - 2LXL^*)$$

$$\theta_1(X) = [L, X], \theta_2(X) = [X, L^*] = \theta_1(X)^*$$

The measurement model is

$$Y(t) = U(t)^*Y_i(t)U(t), Y_i(t) = \bar{c}A(t) + cA(t)^*, c \in \mathbb{C}$$

Quantum Ito's formula is

$$dA.dA^* = dt$$

We have by Quantum Ito's formula,

$$dY(t) = dY_i(t) + j_t(cL + \bar{c}L^*)dt, j_t(Z) = U(t)^*ZU(t)$$

for any Z defined in $\mathfrak{h} \otimes \Gamma_s(L^2(\mathbb{R}_+))$, where $\mathfrak{h}$ is the system Hilbert space and $\Gamma_s(L^2(\mathbb{R}_+))$ is the bath Boson Fock space. Note that $U(t)$ is a unitary operator in $\mathfrak{h} \otimes \Gamma_s(L^2(\mathbb{R}_+))$. In classical filtering theory, the state $X(t)$ evolves according the a classical sde:

$$dX(t) = f(X(t))dt + g(X(t))dB(t)$$

where $B(.)$ is classical Brownian motion and the measurement model is

$$dY(t) = h(X(t))dt + \sigma_V.dV(t)$$

where $V(.)$ Brownian motion independent of $B(.)$. In classical probability therefore, the homomorphism j_t acts on the commutative algebra of real valued functions ϕ on $\mathbb{R}^n$ where $X(t) \in \mathbb{R}^n$ and is defined by

$$j_t(\phi) = \phi(X(t))$$

and it satisfies the sde

$$dj_t(\phi) = \phi'(X(t))dX(t) + (1/2)Tr(gg^T(X(t))phi''(X(t)))dt$$

$$= j_t(K\phi)dt + \sum_{k,m}^{d} B_m(t)j_t(g_{km}(X(t))D_k\phi)$$

where

$$K = f^T D + (1/2)Tr(gg^T.DD^T)$$

is the generator of the Markov process $X(t)$ and $D = \partial/\partial\mathbf{X}$ is the gradient operator. In the scalar process case, there is just one state variable and one Brownian motion $B(.)$ and then the above simplifies to

$$dj_t(\phi) = j_t(K\phi)dt + j_t(gD\phi)dB(t)$$

where

$$K = fD + (1/2)g^2D^2, D = d/dX$$

The measurement model in the classical case can be expressed as

$$dY(t) = j_t(h)dt + \sigma_V dV(t)$$

In the quantum case, the Hermitian operator $cL + \bar{c}L^*$ plays the role of h and $\bar{c}A(t) + cA(t)^*$ plays the role of the classical measurement noise $\sigma_V V(t)$. Note that $\bar{c}A(t) + cA(t)^*$ by itself is a classical Brownian motion process in any state of the system and bath with a variance of $|c|^2$ in place of σ_V^2. The quantum analogue of the classical generator K is the quantum Lindblad generator θ_0. Note that the classical observable ϕ is replaced by the quantum observable X and $\phi(X(t)) = j_t(\phi)$ by $j_t(X) = U(t)^*XU(t)$. $\theta_1(X), \theta_2(X) = \theta_1(X)^*$ are reduced in the classical theory to $gD\phi$ if we take $B(t) = A(t) + A(t)^*$. Note that in the quantum scenario, the process noise $A(t) + A(t)^*$ generally is correlated with the measurement noise $\bar{c}A(t) + cA(t)^*$. It is uncorrelated only when we choose c to be purely imaginary.

Remark: In order to see the classical-quantum analogy better, we must relate the classical theory to classical mechanics, Hamiltonians etc. Thus, we consider the Langevin equation for a classical particle moving in a potential $U(q)$. The stochastic differential equation of motion of such a particle is given by

$$dq(t) = p(t)dt, dp(t) = -(\gamma p(t) + U'(q(t)))dt + g(q(t), p(t))dB(t)$$

The state of the system at time t is now

$$X(t) = [q(t), p(t)]^T$$

and the homomorphism j_t is

$$j_t(\phi) = \phi(X(t))$$

Then, by Ito's formula,

$$dj_t(\phi) = j_t(K\phi) + j_t(S\phi)dB(t)$$

where

$$K\phi = p\partial\phi/\partial q - (\gamma p + U'(q))\partial\phi/\partial p + (1/2)g^2\partial^2\phi/\partial p^2$$

Noting that the Hamiltonian of the particle is

$$H = p^2/2 + U(q)$$

we can write

$$K\phi = \{\phi, H\}_P - \gamma p\partial\phi/\partial p + (1/2)g^2\partial^2\phi/\partial p^2$$

where $\{.,.\}_P$ denotes the Poisson bracket. To see what the quantum generalization of this is, we choose $L = \psi(q, p)$ where now q, p are Hermitian operators satisfying the commutation relations

$$[q, p] = ih/2\pi$$

Then,

$$\theta_0(X) = i[H, X] - (1/2)(LL^*X + XLL^* - 2LXL^*)$$
$$= i[H, X] - (1/2)(L[L^*, X] + [X, L]L^*)$$

Taking

$$X = \phi(q, p)$$

we have

$$i[H, X] = i[p^2/2 + U(q), \phi(q, p)] =$$
$$(i/2)([p, \phi]p + p[p, \phi]) + i[U(q), \phi]$$
$$= (1/2)\{\partial_q\phi, p\} + i[U, \phi]$$

where $\{.,.\}$ denotes the anticommutator. In the classical case where Poisson brackets are replaced by Lie brackets, this expression becomes

$$\{\phi, H\}_P = p\partial_q\phi - (\partial_q U).(\partial_p\phi)$$

but we see that in the quantum case, there are additional factors in view of the non-commutativity of q and p that are expressible as a power series in Planck's constant. For example, suppose

$$\phi(q, p) = \phi_0(q) + \{\phi_1(q), p\}$$

then we get

$$i[H,X] = (1/2)\{\partial_q\phi_0(q),p\} + (1/2)\{\{\partial_q\phi_1(q),p\},p\}$$

In the classical case, these equations simplify as

$$\phi(q,p) = \phi_0(q) + 2p\phi_1(q),$$

and $i[H,X]$ gets replaced by

$$p\partial_q\phi_0(q) + 2p^2\partial_q\phi_1(q)$$

A further special case of this is when

$$X = \phi(q,p) = \phi_0(q)$$

Then,

$$i[H,X] = (1/2)\{\partial_q\phi_0(q),p\}$$

while in the classical case, this reduces to

$$p\partial_q\phi_0(q)$$

As for the Lindblad terms, in the quantum case with $X = \phi(q,p)$ and $L = \psi(q,p)$, we find that the quantum version of

$$-\gamma p\partial\phi/\partial p + (1/2)g^2\partial^2\phi/\partial p^2$$

should be simply

$$-(1/2)(LL^*X + XLL^* - 2LXL^*) = (-1/2)(L[L^*,X] + [X,L]L^*)$$

where

$$X = \phi(q,p)$$

and $L = \psi(q,p)$ chosen appropriately in terms of the classical function $g(q,p)$. For example, choosing

$$L = aq + bp, a, b \in \mathbb{C},$$

we get

$$[L^*,X] = [\bar{a}q + \bar{b}p, \phi(q,p)] = i\bar{a}\partial_p\phi - i\bar{b}\partial_q\phi$$

$$[X,L] = -ia\partial_p\phi + ib\partial_q\phi$$

$$\begin{aligned} L[L^*,X] + [X,L]L^* &= (aq+bp)(i\bar{a}\partial_p\phi - i\bar{b}\partial_q\phi) \\ &+(-ia\partial_p\phi + ib\partial_q\phi)(\bar{a}q + \bar{b}p) \\ &= i|a|^2[q,\partial_p\phi] + ib\bar{a}(p.\partial_p\phi + \partial_q\phi.q) \\ &-ia\bar{b}(q\partial_q\phi + \partial_p\phi.p) - i|b|^2[p,\partial_q\phi] \\ &= -|a|^2\partial_p^2\phi - |b|^2\partial_q^2\phi \end{aligned}$$

$$i b\bar{a}(p.\partial_p\phi + \partial_q\phi.q) - ia\bar{b}(q\partial_q\phi + \partial_p\phi.p)$$

(Remark: To get some sort of agreement with the classical case, the term involving ∂_q^2 should not appear. Thus, we set $b = 0$, ie,

$$L = aq$$

in which case,

$$L[L^*, X] + [X, L]L^* = -|a|^2\partial_p^2\phi$$

but then we are not able to get the damping term $\gamma.p\partial_p\phi$). So we see that if we constrain the Lindblad operator L to be linear in q and p, we are able to obtain some sort of a quantum analogue of the classical Langevin equation but with some additional terms. Moreover, by restricting L to be linear in q, p, we cannot in the quantum case account for a general diffusion coefficient $g^2(q,p)$ dependent on q, p present in the classical case.

Upto this, we have dealt with drawing analogies between the classical Fokker-Planck equation for stochastic differential equations in classical mechanics and probability on the one hand and quantum stochastic differential equations in quantum mehcanics and quantum probability on the other. Now, we try to draw analogies between the classical filtering and quantum filtering equations.

First we note that in the quantum case, the measurement noise process $Y_i(t) = \bar{c}A(t) + cA(t)^*$ is also a Brownian motion with a variance parameter $|c|^2$ and the process noise in $dj_t(X)$ appears as $j_t([L, X])dA(t) + j_t([X, L^*])dA(t)^*$. This measurement noise is generally correlated with the process noise unless it happens that $L^* = L$ and c is real, in which case we get that the process noise appears as $j_t([L, X])(dA(t) - dA(t)^*)$ with $[L, X]^* = [X, L] = -[L, X]$ while the measurement noise differential is $c(dA(t) + dA(t)^*)$. Another case in which this happens is when c is pure imaginary and $L^* = -L$ in which case, it happens that the process noise appears as $j_t([L, X])(dA(t) + dA(t)^*)$ (since $[L, X]^* = [X, L^*] = -[X, L] = [L, X]$) while the measurement noise differential is $\bar{c}(dA(t) - dA(t)^*)$. These two cases are the only cases in which the process noise and measurement noise are independent Brownian motions as in the classical model for nonlinear filtering.

The Belavkin filter is obtained by denoting the non-demolition measurements upto time t by

$$Z_t = \sigma(Y(s) : s \leq t)$$

and writing

$$\mathbb{E}(j_t(X)|Z_t) = \pi_t(X)$$

and noting that $\pi_t(X), t \geq 0$ form an Abelian family of operators along with Z_t and hence we can assume that they satisfy an equation

$$d\pi_t(X) = F_t(X)dt + G_t(X)dY(t)$$

where $F_t(X), G_t(X)$ are Z_t measurable. Then assuming that

$$dC(t) = f(t)C(t)dY(t), C(0) = 1$$

we get that $C(t)$ is Z_t-measurable and therefore by the basic orthogonality principle in signal estimation theory,

$$\mathbb{E}[(j_t(X) - \pi_t(X))C(t)] = 0$$

[16] Simultaneous application of the representation theory of the permutation groups and the Euclidean motion group in three dimensions to 3-D image processing problems.

Assume that we are given n $3-D$ objects whose centres are located at the positions $\mathbf{r}_j, j = 1, 2, ..., n$. The kth object whose centre is located at $\mathbf{r}_k$ will emit a signal $f_k(\mathbf{r} - \mathbf{r}_k)$. Thus, the total signal emitted by all the n objects is given by

$$X(\mathbf{r}|\mathbf{r}_1, ..., \mathbf{r}_n) =) = \sum_{k=1}^{n} f_k(\mathbf{r} - \mathbf{r}_k)$$

Now suppose we permute the objects by applying a permutation $\sigma \in S_n$ and also rotate the entire array of objects by the rotation $\mathbf{R} \in SO(3)$. Then the resulting signal field becomes

$$Y(\mathbf{r}|\mathbf{r}_1, ..., \mathbf{r}_k) = \sum_{k=1}^{n} f_k(\mathbf{R}^{-1}\mathbf{r} - \mathbf{r}_{\sigma^{-1}k})$$

$$= X(\mathbf{R}^{-1}\mathbf{r}|\mathbf{r}_{\sigma^{-1}1}, ..., \mathbf{r}_{\sigma^{-1}n})$$

From measurements of the signal fields $\mathbf{X}$ and $\mathbf{Y}$, I wish to estimate the rotation $\mathbf{R}$ and the permutation σ. Let π be a unitary representation of $SO(3)$ and η a unitary representation of S_n. We compute

$$\sum_{\rho \in S_n} Y(\mathbf{r}|\mathbf{r}_{\rho 1}, ..., \mathbf{r}_{\rho n})\eta(\rho)$$

$$= \sum_{\rho \in S_n} X(\mathbf{R}^{-1}\mathbf{r}|\mathbf{r}_{\sigma^{-1}\rho 1}, ..., \mathbf{r}_{\sigma^{-1}\rho n})\eta(\rho)^*$$

$$= [\sum_{\rho \in S_n} X(\mathbf{R}^{-1}\mathbf{r}|\mathbf{r}_{\rho 1}, ..., \mathbf{r}_{\rho n})\eta(\rho)^*]\eta(\sigma)^*$$

Also,

$$\int_{SO(3)} Y(\mathbf{Sr}|\mathbf{r}_1, ..., \mathbf{r}_n)\pi(\mathbf{S})^* d\mathbf{S}$$

$$= \int X(\mathbf{R}^{-1}\mathbf{Sr}|\mathbf{r}_{\sigma^{-1}1}, ..., \mathbf{r}_{\sigma^{-1}n})\pi(\mathbf{S})^* d\mathbf{S}$$

$$= \int X(\mathbf{Sr}|\mathbf{r}_{\sigma^{-1}1}, ..., \mathbf{r}_{\sigma^{-1}n})\pi(\mathbf{RS})^* d\mathbf{S}$$

$$= [\int X(\mathbf{Sr}|\mathbf{r}_{\sigma^{-1}1}, ..., \mathbf{r}_{\sigma^{-1}n})\pi(\mathbf{S})^* d\mathbf{S}]\pi(\mathbf{R})^*$$

More generally, we then have

$$\sum_{\rho \in S_n} \int Y(\mathbf{Sr}|\mathbf{r}_{\rho 1}, ..., \mathbf{r}_{\rho n})\pi(\mathbf{S})^* \otimes \eta(\rho)^* d\mathbf{S}$$

$$= [\sum_{\rho \in S_n} \int X(\mathbf{Sr}|\mathbf{r}_{\rho 1}, ..., \mathbf{r}_{\rho n})\pi(\mathbf{S})^* \otimes \eta(\rho)^* d\mathbf{S}](\pi(\mathbf{R})^* \otimes \eta(\sigma)^*)$$

This equation gives us a clue to how linear estimation theory can be applied to estimate both $\mathbf{R} \in SO(3)$ and $\sigma \in S_n$.

[17] (Part of a B.E. project) In quantum mechanics, probabilities of events are computed w.r.t a given state of the system. If the system is in a pure state $|\psi>$, then the probability of an event descrribed by the projection P is given by $<\psi|P|\psi>$. Many times, calculating the pure state wave function $|\psi>$ becomes very complicated because we have to solve a Schrodinger equation for that. A typical example is that of a two electron atom like Helium where the wave function is a function of two position variables, ie, a function of six real variables. However, there do exist approximate ways of calculating the required wave function. One such is the Hartree-Fock method in which for example in a two electron atom, we know that the wave function must be antisymmetric w.r.t interchange of the two position and spin variables owing to the Pauli exclusion principle which states that two electrons cannot occupy the same state, ie, have the same positions and spins. The Hartree-Fock approximation involves assuming that the wave function is a product of position wave functions and spin wave functions with one of them being symmetric and the other one antisymmetric w.r.t interachange of the two electrons. Further, we assume that the position part of the wave function if antisymmetric can be represented as the antisymmetrizer of the product of single particle position wave functions and if symmetric, can be represented as the symmetrizer of the product of single particle position wave functions with the same argument being valid for the spin wave functions.

I outline here most of the steps involved in doing the Hartree-Fock simulation for a two electron atom with interacting spins and angular momenta. The Hamiltonian without taking spin or orbital momentum interactions is given by

$$H_{01} + H_{02} + V_{12}$$

Where

$$H_{01} = p_1^2/2m - 2e^2/r_1, H_{02} = p_2^2/2m - 2e`2/r_2 V_{12} = e^2/|r_2 - r_1|$$

Where

$$p_1 = -i\nabla_1, p_2 = -i\nabla_2$$

The magnetic field produced by the two nuclear protons at the site of the first electron in view of their relative motions is given by

$$B_{p1}(r_1) = ev_1 \times r_1/r_1^3 = c_1 L_1/r_1^3, L_1 = r_1 \times p_1$$

The interaction energy between this magnetic field and the total spin orbital magnetic moment

$$M_1 = (-e/2m)(L_1 + g\sigma_1)$$

Of the first electron is given by

$$V_{p1} = -(M_1, B_{p1}(r_1)) = c_1(L_1 + g\sigma_1, L_1)/r_1^3$$

Likewise the magnetic interactions between the proton and the total magnetic moment of the second electron is given by

$$V_{p2} = c_2(L_2 + g\sigma_2, L_2)/r_2^3$$

Now the first electron moving with a velocity of v_1 produces at the site of the second electron moving with a relative velocity of $v_2 - v_1$ w.r.t the first electron, a magnetic field at the site of the second electron given by

$$B_{m21} = -e(v_1 - v_2)x(r_2 - r_1)/|r_2 - r_1|^3 = (e/m)(-L_1 - L_2 - p_1 x r_2 - p_2 x r_1)/|r_2 - r_1|^3$$

Since

$$V_1 = p_1/m, v_2 = p_2/m, L_1 = r_1 x p_1, L_2 = r_2 x p_2$$

and further the spin magnetic moment of the first electron produces a magnetic field

$$B_{s21} = curl_2((-ge\sigma_1/2m)x(r_2 - r_1)/|r_2 - r_1|^3)$$

Remark: A magnetic moment m at the origin produces a magnetic vector potential

$$A = m \times r/r^3$$

And a magnetic field

$$B = curl A = curl(m \times r/r^3) = -curl(m \times \nabla(1/r))$$

$$= -m\nabla^2(1/r) + (m, \nabla)]\nabla(1/r)$$

$$= (m, \nabla)((-\mathbf{r}/r^3) = -m_x(\partial_x(\mathbf{r}/r^3)) - m_y\partial_y(\mathbf{r}/r^3) - m_z\partial_z(r/r^3)$$

$$= -m_x(\hat{x}/r^3 - 3\mathbf{r}x/r^5) - m_y((hatx/r^3 - 3\mathbf{r}y/r^5) - m_z(\hat{x}/r^3 - 3\mathbf{r}z/r^5)$$

$$= -\mathbf{m}/r^3 + 3(\mathbf{m}, \mathbf{r})\mathbf{r}/r^5$$

Thus the interaction energy between the total magnetic moment of the second electron with the magnetic field generated by the first electron due to its relative motion and spin is given by

$$(B_{m21} + B_{s21}, (e/2m)(L_2 + g\sigma_2))$$

$$= (e/2m)(L_2 + g\sigma_2, (-L_1 - L_2 - p_1 x r_2 - p_2 x r_1))/|r_2 - r_1|^3$$
$$+(ge/2m)(L_2 + g\sigma_2, \sigma_1/|r_2 - r_1|^3 - 3(\sigma_1, r_2 - r_1)(\mathbf{r_2} - \mathbf{r_1})/|r_2 - r_1|^5)$$

Likewise, the interaction energy between the total magnetic moment of the first electron with the magnetic moment of the first electron and the magnetic field generated by the second electron due to its relative motion and spin is given by

$$(B_{m12} + B_{s12}, (e/2m)(L_1 + g\sigma_1))$$

$$(e/2m)(L_1 + g\sigma_1, (-L_1 - L_2 - p_1 x r_2 - p_2 x r_1))/|r_2 - r_1|^3$$
$$+(ge/2m)(L_1 + g\sigma_1, \sigma_2/|r_2 - r_1|^3 + 3(\sigma_2, r_2 - r_1)(\mathbf{r_2} - \mathbf{r_1})/|r_2 - r_1|^5)$$

Remark: There are other ways to calculate the interaction energy between the two electrons involving magnetic fields produced by them and their magnetic moments. They give different results. For example, the magnetic field produced by the first electron based on its motion and spin magnetic moment is given by

$$B_1(r) = -ev_1 \times (r - r_1)/|r - r_1|^3 + curl(m_1 \times (r - r_1))/|r - r_1|^3)$$

$$= -ev_1 \times (r - r_1)/|r - r_1|^3 - \mathbf{m}_1/|r - r_1|^3 + 3(\mathbf{m_1}, \mathbf{r} - \mathbf{r_1})\mathbf{r} - \mathbf{r_1}/|r - r_1|^5$$

Where

$$\mathbf{m}_1 = -ge\sigma_1/2m$$

Is the spin magnetic moment of the first electron. Likewise, the magnetic field produced by the second electron is

$$B_2(r) = -ev_2 \times (r - r_2)/|r - r_2|^3 + curl(m_2 \times (r - r_2))/|r - r_2|^3)$$

$$= -ev_2 \times (r - r_2)/|r - r_2|^3 - \mathbf{m}_2/|r - r_2|^3 + 3(\mathbf{m_2}, \mathbf{r} - \mathbf{r_2})\mathbf{r} - \mathbf{r_2}/|r - r_2|^5$$

The total energy in the magnetic field produced by the two electrons is

$$E_B = (\mu/2)\int |B_1(r) + B_2(r)|^2 d^3r$$

And the interaction part of this energy is clearly

$$(\mu/2)\int (B_1(r), B_2(r))d^3r$$

From the above consideration, it is clear that the magnetic interaction energy must be a scalar operator built out of the vector operators $\mathbf{p_1}, \mathbf{p_2}, \mathbf{r_2} - \mathbf{r_1}, \sigma_1, \sigma_2)$. It is then easy to see that this interaction energy must have the form

$$f_1(|r_2 - r_1|)(p_1, p_2) + f_2(|r_2 - r_1|)(p_1, \sigma_2) + f_2(|r_2 - r_1|)(p_2, \sigma_1) + f_3(|r_2 - r_1|)((r_2 - r_1)$$
$$\times p_1, \sigma_2) + f_3(|r_2 - r_1|)((r_1 - r_2) \times p_2, \sigma_1) + f_4(|r_2 - r_1|)(\sigma_1, \sigma_2) + f_5(|r_2 - r_1|)(\sigma_1 \times \sigma_2, r_2 - r_1)$$

In order to formulate the Hartree-Fock equations taking spin into account, we must therefore First discretize the spatial region into N^3 pixels, N being the number of pixels along each of the xyz coordinate axes. Then $p_{1x}, p_{1y}, p_{1z}, x_1, y_1, z_1$ are each represented by $N^3 \times N^3$ matrices, ie, these act in the Hilbert space

$\mathbb{C}^{N^3\times 1}$. Likewise For $p_{2x}, p_{2y}, p_{2z}, x_2, y_2, z_2$. These act in another independent Hilbert space $\mathbb{C}^{N^3}$ Each of the spin matrices $\sigma_{1x}, \sigma_{1y}, \sigma_{1z}$ is represented by a 2×2 Hermitian matrix, ie, these act in the Hilbert space $\mathbb{C}^2$. Likewise, $\sigma_{2x}, \sigma_{2y}, \sigma_{2z}$ act in another independent Hilbert space $\mathbb{C}^2$. In this The total of all the above operators therefore acts in the tensor product Hilbert space

$$\mathcal{H} = \mathbb{C}^{N^3} \otimes \mathbb{C}^{N^3} \otimes \mathbb{C}^2 \otimes \mathbb{C}^2 = \mathbb{C}^{4N^6}$$

The Hamiltonian of the system thus has the following decomposition:

$$H = H_1 \otimes I_{3N^3} \otimes I_2 \otimes I_2 + I_{3N^3} \otimes H_2 \otimes I_2 \otimes I_2$$
$$+V$$

Where V acts in the joint tensor product space and can be decomposed as

$$V = \sum_{k=1}^{p} (V_{1k} \otimes V_{2k} \otimes V_{3k} \otimes V_{4k})$$

Where V_{2k} is a function of only the position and momentum variables of the first electron, and thus acts in $\mathbb{C}^{3N^3}$, namely the first tensor product component in $\mathcal{H}$. V_{2k} is a function of only the position and momentum variables of the second electron and thus acts in the second tensor product component $\mathbb{C}^{3N^3}$ of $\mathcal{H}$. V_{3k} is a function of only the spin matrices of the first electron and therefore acts in the third tensor product component $\mathbb{C}^2$ and finally, V_{4k} is a function of only the spin matrices of the second electron and therefore acts in the fourth tensor product component $\mathbb{C}^2$. Accordingly, in the Hartree Fock approximation, we can assume that the state ψ of the two electrons has one of the following forms:
[a]

$$C(\psi_1 \otimes \psi_2 - \psi_2 \otimes \psi_1)| ++ >$$

[b]

$$C(\psi_1 \otimes \psi_2 - \psi_2 \otimes \psi_1)| -- >$$

[c]

$$C(\psi_1 \otimes \psi_2 - \psi_2 \otimes \psi_1)(| +- > + | -+ >)/\sqrt{2}$$

[d]

$$C(\psi_1 \otimes \psi_2 + \psi_2 \otimes \psi_1)(| +- > - | -+ >)/\sqrt{2}$$

Where ψ_1 is a vector in the first tensor product component, ψ_2 in the second, and objects such as $|++>, |-->, |+->, |-+>$ have have their first component in the third tensor product components and their second component in the fourth tensor product component. Substituting for example the first expression [a] for ψ gives us on noting that H_1 and H_2 are identical matrices so are V_{1k} and V_{2k} and so also are V_{3k} and V_{4k} for each k the following expression for the average of H in the state $|\psi>$: [a]

$$<\psi|H|\psi>=$$
$$2C^2[<\psi_1|H_1|\psi_1> + <\psi_2|H_1|\psi_2> - <\psi_1|H_1|\psi_2><\psi_2|\psi_1>$$
$$-<\psi_2|H_1|\psi_1><\psi_1|\psi_2>]$$

$$+C^2\sum_{k=1}^{p} <\psi_1\otimes\psi_2-\psi_2\otimes\psi_1|V_{1k}\otimes V_{2k}|\psi_1\otimes\psi_2-\psi_2\otimes\psi_1><++|V_{3k}\otimes V_{4k}|++>$$

$$=2C^2[<\psi_1|H_1|\psi_1>+<\psi_2|H_1|\psi_2>-<\psi_1|H_1|\psi_2><\psi_2|\psi_1>-<\psi_2|H_1|\psi_1><\psi_1|\psi_2>]$$

$$+2C^2\sum_{k=1}^{p}[(<\psi_1|V_{1k}|\psi_1><\psi_2|V_{1k}|\psi_2>-<\psi_1|V_{1k}|\psi_2><\psi_2|V_{1k}|\psi_1>)|<+|V_{3k}|+>|^2]$$

We are putting the normalization constraints:

$$<\psi_1|\psi_1>=1=<\psi_2|\psi_2>,$$

$$2C^2(1-|<\psi_1|\psi_2>|^2)=1$$

We must now apply the variational principle to extremize $<\psi|H|\psi>$ w.r.t $|\psi_1>$ and $|\psi_2>$ subject to the above constraints.

Likewise, for the second case [b]

$$<\psi|H|\psi>=$$

$$2C^2[<\psi_1|H_1|\psi_1>+<\psi_2|H_1|\psi_2>-<\psi_1|H_1|\psi_2><\psi_2|\psi_1>-<\psi_2|H_1|\psi_1><\psi_1|\psi_2>]$$

$$+2C^2\sum_{k=1}^{p}[(<\psi_1|V_{1k}|\psi_1><\psi_2|V_{1k}|\psi_2>-<\psi_1|V_{1k}|\psi_2><\psi_2|V_{1k}|\psi_1>)|<-|V_{3k}|->|^2]$$

For the third case, [c]

$$<\psi|H|\psi>=$$

$$2C^2[<\psi_1|H_1|\psi_1>+<\psi_2|H_1|\psi_2>-<\psi_1|H_1|\psi_2><\psi_2|\psi_1>$$
$$-<\psi_2|H_1|\psi_1><\psi_1|\psi_2>]$$

$$+2C^2\sum_{k=1}^{p}[(<\psi_1|V_{1k}|\psi_1><\psi_2|V_{1k}|\psi_2>-<\psi_1|V_{1k}|\psi_2><\psi_2|V_{1k}|\psi_1>$$
$$\times(1/2)(<+-|+<-+|)(V_{3k}\otimes V_{4k})(|+->+|-+>)$$

Now observe that

$$(<+-|+<-+|)(V_{3k}\otimes V_{4k})(|+->+|-+>)=$$

$$<+-|V_{3k}\otimes V_{4k}|+->+<-+|V_{3k}\otimes V_{4k}|-+>$$
$$+<+-|V_{3k}\otimes V_{4k}|-+>+<-+|V_{3k}\otimes V_{4k}|+->$$

and finally for the fourth case, [d]

$$<\psi|H|\psi>=$$

$$2C^2[<\psi_1|H_1|\psi_1>+<\psi_2|H_1|\psi_2>+<\psi_1|H_1|\psi_2><\psi_2|\psi_1>$$
$$+<\psi_2|H_1|\psi_1><\psi_1|\psi_2>]+$$

$$2C^2\sum_k [<\psi_1|V_{1k}|\psi_1><\psi_2|V_{1k}|\psi_2> + <\psi_1|V_{1k}|\psi_2><\psi_2|V_{1k}|\psi_1>]$$
$$\times(1/2)(<+-|-<-+|)(V_{3k}\otimes V_{4k})(|+->-|-+>)$$

We work out the details for case [a]. Assume that the overlap constant C^2 is fixed and then after taking into account normalization constraints using Lagrange multipliers, the functional to be extremized is

$$S=<\psi|H|\psi> -E_1(<\psi_1|\psi_1>-1)-E_2(<\psi_2|\psi_2>-1)$$
$$-E_{12}(1-1/2C^2-|<\psi_1|\psi_2>|^2)$$

Setting the variational derivative

$$\delta S/\delta\psi_1^* = 0$$

here, gives us

$$2C^2[H_1|\psi_1> - <\psi_2|\psi_1> H_1|\psi_2> - <\psi_2|H_1|\psi_1>|\psi_2>]$$
$$+2C^2\sum_k [|<+|V_{3k}|+>|^2(<\psi_2|V_{1k}|\psi_2>V_{1k}|\psi_1> - <\psi_2|V_{1k}|\psi_1>V_{1k}|\psi_2>)$$
$$-E_1|\psi_1> +E_{12}<\psi_2|\psi_1>|\psi_2>=0$$

Setting the variational derivative

$$\delta S/\delta\psi_2^* = 0$$

gives us

$$2C^2[H_1|\psi_2> - <\psi_1|H_1|\psi_2>|\psi_2> - <\psi_1|\psi_2>H_1|\psi_1>]$$
$$+2C^2\sum_k [|<+|V_{1k}|+>|^2(<\psi_1|V_{1k}|\psi_1>V_{1k}|\psi_2> - <\psi_1|V_{1k}|\psi_2>V_{1k}|\psi_1>)$$
$$-E_2|\psi_2> +E_{12}<\psi_1|\psi_2>|\psi_1>=0$$

If we put the restriction of no overlap, ie

$$<\psi_2|\psi_1>=0$$

then we get $C^2 = 1/2$ and the above equations simplify to

$$[H_1|\psi_1> - <\psi_2|H_1|\psi_1>|\psi_2>]$$
$$+\sum_k [|<+|V_{3k}|+>|^2(<\psi_2|V_{1k}|\psi_2>V_{1k}|\psi_1>$$
$$- <\psi_2|V_{1k}|\psi_1>V_{1k}|\psi_2>)-E_1|\psi_1>=0---(a)$$

$$[H_1|\psi_2> - <\psi_1|H_1|\psi_2>|\psi_2>]$$
$$+\sum_k [|<+|V_{1k}|+>|^2(<\psi_1|V_{1k}|\psi_1>V_{1k}|\psi_2>$$
$$- <\psi_1|V_{1k}|\psi_2>V_{1k}|\psi_1>)-E_2|\psi_2>=0---(b)$$

Note that (a) and (b) are consistent with the no overlap requirement $<\psi_2|\psi_1>=$ 0 as follows by premultiplying (a) by $<\psi_2|$ and (b) by $<\psi_1|$ and obtaining an identity. It is clear that in the time dependent case, we have to replace the energy values E_1 and E_2 by $i\partial/\partial t$. This is a consequence of the fact that under the no overlap condition,

$$<\psi_1\otimes\psi_2-\psi_2\otimes\psi_1|\partial_t|\psi_1\otimes\psi_2-\psi_2\otimes\psi_1>=$$

$$<\psi_1|\partial_t|\psi_1>+<\psi_2|\partial_t|\psi_2>$$

since we are assuming that $|\psi_k>, k=1,2$ are normalized.

Remarks: $|\psi>$ is the overall wave function of the positions and spins of the two electrons. The forms of $|\psi>$ assumed in [a], [b], [c], [d] ensure that it is antisymmetric w.r.t the interchange of the positions and spins of the two electrons. For example, in [a], [b] and [c], the wave function is antisymmetric w.r.t. the interchange of the positions of the two electrons and symmetric w.r.t the interchange of the spin of two two electrons. Hence, since the product of a minus and a plus is a minus, the overall wave function is antisymmetric w.r.t the interchange of both the positions and spins of the two electrons. In [d], the wave functions is symmetric w.r.t the interchange of the positions of the two electrons while it is antisymmetric w.r.t the interchange of the two electron spins so once again the overall wave functions is antisymmetric. We could represent such wave functions alternately as $\psi(\mathbf{r}_1,s_1,\mathbf{r}_2,s_2)$ where $\mathbf{r}_k$, and s_k are respectively tho position and z-spin component of the kth electron $k=1,2$. Thus in the case [a],

$$\psi(\mathbf{r}_1,1/2,\mathbf{r}_2,1/2)=C(\psi_1\otimes\psi_2-\psi_2\otimes\psi_1)(r_1,r_2)$$
$$=C(\psi_1(r_1)\psi_2(r_2)-\psi_1(r_2)\psi_2(r_1))$$

and $\psi(\mathbf{r}_1,s_1,\mathbf{r}_2,s_2)$ is zero for the choices $(1/2,-1/2,),(-1/2,1/2),(-1/2,-1/2)$ for (s_1,s_2). We can treat this wave function as a four component wave function of the position variables. Likewise, for the wave function of [d], it equals $C(\psi_1(r_1)\psi_2(r_2)+\psi_1(r_2)\psi_2(r_1))/\sqrt{2}$ for the choice $(s_1,s_2)=(1/2,-1/2)$, the negative of the same for $(s_1,s_2)=(-1/2,1/2)$ and zero for the remaining two choices of (s_1,s_2). The antisymmetry of these wave functions are therefore equivalently expressed as

$$\psi(\mathbf{r}_1,s_1,\mathbf{r}_2,s_2)=-\psi(\mathbf{r}_2,s_2,\mathbf{r}_1,s_1),\mathbf{r}_1,\mathbf{r}_2\in\mathbb{R}^3,s_1,s_2=\pm 1/2$$

Now we come to the point of how to express the spin-orbit interaction terms in the form $\sum_k V_{1k}\otimes V_{2k}\otimes V_{3k}\otimes V_{4k}$. The general form of this interaction terms, as discussed above is of the form

$$(\mathbf{f}_1(\mathbf{r}_1,\mathbf{r}_2,\mathbf{p}_1,\mathbf{p}_2,\sigma_1)$$

$$+(\mathbf{f}_1(\mathbf{r}_1,\mathbf{r}_2,\mathbf{p}_1,\mathbf{p}_2,\sigma_2)$$

where

$$(\mathbf{f}_1(\mathbf{r}_1,\mathbf{r}_2,\mathbf{p}_1,\mathbf{p}_2,\sigma_1)$$

$$= f_{1x}(\mathbf{r}_1, \mathbf{r}_2, \mathbf{p}_1, \mathbf{p}_2)\sigma_{1x} + f_{1y}(\mathbf{r}_1, \mathbf{r}_2, \mathbf{p}_1, \mathbf{p}_2)\sigma_{1y} + f_{1z}(\mathbf{r}_1, \mathbf{r}_2, \mathbf{p}_1, \mathbf{p}_2)\sigma_{1z}$$

and likewise for the second term. The x, y, z components of the vector operators $\mathbf{r}_1, \mathbf{p}_1$ act in the Hilbert space $\mathcal{H}_1 = L^2(\mathbb{R}^3)$ while the components of $\mathbf{r}_2, \mathbf{p}_2$ act in another independent and identical Hilbert space $\mathcal{H}_2 = L^2(\mathbb{R}^3)$. The x,y,z components of σ_1 act in another independent Hilbert space $\mathbb{C}^2$ and finally, the x, y, z components of σ_2 act in yet another independent Hilbert space $\mathcal{H}_4 = \mathbb{C}^2$. All the components of the Hamiltonian then act in the tensor product Hilbert space

$$\mathcal{H} = \mathcal{H}_1 \otimes \mathcal{H}_2 \otimes \mathcal{H}_3 \otimes \mathcal{H}_4$$

After discretization, V_{1k}, V_{2k} become $N^3 \times N^3$ matrices while V_{3k}, V_{4k} are 2×2 matrices. $\mathcal{H}$ becomes the finite dimensional Hilbert space $\mathbb{C}^{4N^6} = \mathbb{C}^{N^3} \otimes \mathbb{C}^{N^3} \otimes \mathbb{C}^2 \otimes \mathbb{C}^2$. For example, consider an interaction term of the form

$$f_1(x_1, y_1, z_1) f_2(x_2, y_2, z_2) p_{1x}\sigma_{1x} = -i f_1(x_1, y_1, z_1) f_2(x_2, y_2, z_2)\sigma_{1x}\partial/\partial x_1$$

We require to represent it in the form $\sum_k V_{1k} \otimes V_{2k} \otimes V_{3k} \otimes V_{4k}$. Let $e(k)$ denote the $N \times 1$ column vector with a one in the kth position and zeros at all the other positions where $k = 1, 2, ..., N$. Likewise, let $f(k)$ denote the 2×1 vector with a one in the kth position and zero at the other position $k = 1, 2$. Then $-i\partial/\partial x_1$ is represented by an $N^3 \times N^3$ matrix that takes the vector

$$\phi = \sum_{n_{1x}, n_{1y}, n_{1z}=1}^{N} \phi(n_{1x}, n_{1y}, n_{1z}) e(n_{1x}) \otimes e(n_{1y}) \otimes e(n_{1z}) \in \mathcal{H}_1 = \mathbb{C}^{N^3}$$

to the vector

$$D_{p_1x}\phi = \sum_{n_{1x}, n_{1y}, n_{1z}} \Delta^{-1}(\phi(n_{1x}+1, n_{1y}, n_{1z}) - \phi(n_{1x}, n_{1y}, n_{1z})) e(n_{1x}) \otimes e(n_{1y}) \otimes e(n_{1z})$$

$$= \sum_{n_{1x}, n_{1y}, n_{1z}} [e(n_{1x}) \otimes e(n_{1y}) \otimes e(n_{1z})].[(e(n_{1x}+1) - e(n_{1x}) \otimes e(n_{1y}) \otimes e(n_{1z})]^T \phi$$

because, we obviously have

$$[e(n_{1x}) \otimes e(n_{1y}) \otimes e(n_{1z})]^T \phi = \phi(n_{1x}, n_{1y}, n_{1z})$$

In other words, D_{1x} is the $N^3 \times N^3$ matrix given by

$$D_{p_1x} = \Delta^{-1} \sum_{n_{1x}, n_{1y}, n_{1z}} [e(n_{1x}) \otimes e(n_{1y}) \otimes e(n_{1z})].[(e(n_{1x}+1) - e(n_{1x})) \otimes e(n_{1y}) \otimes e(n_{1z})]^T$$

Multiplication by $f_1(x_1, y_1, z_1)$ is represented by the $N^3 \times N^3$ diagonal matrix

$$D_{f_1} = \sum_{n_{1x}, n_{1y}, n_{1z}=1}^{N} f_1(n_{1x}\Delta, n_{1y}\Delta, n_{1z}\Delta)[e(n_{1x}) \otimes e(n_{1y}) \otimes e(n_{1z})].[e(n_{1x}) \otimes e(n_{1y}) \otimes e(n_{1z})]^T$$

Likewise $f_2(x_2, y_2, z_2)$ is represented by another $N^3 \times N^3$ diagonal matrix D_{f_2}. Hence,

$$f_1(x_1, y_1, z_1)f_2(x_2, y_2, z_2)p_{1x}\sigma_{1x} = f_1(x_1, y_1, z_1)p_{1x}f_2(x_2, y_2, z_2)\sigma_{1x}$$

will then be represented by the $4N^3 \times 4N^3$ matrix

$$(D_{f_1}D_{p_1}) \otimes D_{f_2} \otimes \sigma_{1x} \otimes I_2$$

which is of the desired form $V_{1k} \otimes V_{2k} \otimes V_{3k} \otimes V_{4k}$. It should be noted that there are terms in the above interaction that cannot be expressed in such a completely factorized form, for example a term involving $1/|r_1 - r_2|$. However such terms generally have the form

$$f(x_1, y_1, z_1, x_2, y_2, z_2)p_{1x}\sigma_{1x} --- (\alpha)$$

or

$$f(x_1, y_1, z_1, x_2, y_2, z_2)p_{1y}\sigma_{1x}$$

or

$$f(x_1, y_1, z_1, x_2, y_2, z_2)p_{1x}\sigma_{1y}$$

etc. and these can be expressed in a restricted factorized form

$$V_{12k} \otimes V_{3k} \otimes V_{4k}$$

where V_{12k} is $N^6 \times N^6$ and acts in $\mathcal{H}_1 \otimes \mathcal{H}_2$ while V_{3k}, V_{4k} are both 2×2 and each act in $\mathbb{C}^2$. In the above case (α),

$$V_{3k} = \sum_{n_{1x}, n_{1y}, n_{1z}, n_{2x}, n_{2y}, n_{2z}=1}^{N} f(n_{1x}\Delta, ..., n_{2z}\Delta)[e(n_{1x}) \otimes e(n_{1y}) \otimes e(n_{1z}) \otimes e(n_{2x} \otimes e(n_{2y}) \otimes e(n_{2z})]['']^T$$

where $['']$ denotes $[e(n_{1x}) \otimes e(n_{1y}) \otimes e(n_{1z}) \otimes e(n_{2x} \otimes e(n_{2y}) \otimes e(n_{2z})]$. When such a term is present in the Hamiltonian, it contributes an amount

$$<\psi|V_{12} \otimes V_3 \otimes V_4|\psi>$$

to $<\psi|H|\psi>$ and for example in case (a), this evaluates to

$$C^2 < (\psi_1 \otimes \psi_2 - \psi_2 \otimes \psi_1)|V_{12}|\psi_1 \otimes \psi_2 - \psi_2 \otimes \psi_1 >< +|V_3|+ >< +|V_4|+ >$$

Now,

$$< (\psi_1 \otimes \psi_2 \quad \psi_2 \otimes \psi_1)|V_{12}|\psi_1 \otimes \psi_2 - \psi_2 \otimes \psi_1 >$$

$$=< \psi_1 \otimes \psi_2|V_{12}|\psi_1 \otimes \psi_2 > + < \psi_2 \otimes \psi_1|V_{12}|\psi_2 \otimes \psi_1 > - < \psi_1 \otimes \psi_2|V_{12}|\psi_2 \otimes \psi_1 >$$

$$- < \psi_2 \otimes \psi_1|V_{12}|\psi_1 \otimes \psi_2 >$$

The variational derivative of this w.r.t ψ_1^* evaluates to

$$(I_{N^3} \otimes \psi_2^*)V_{12}|\psi_1 \otimes \psi_2 > +(\psi_2^* \otimes I_{N^3})V_{12}|\psi_2 \otimes \psi_1 >$$

$$-(I_{N^3} \otimes \psi_2^*)V_{12}|\psi_2 \otimes \psi_1> -(\psi_2^* \otimes I_{N^3})V_{12}|\psi_1 \otimes \psi_2>$$

multiplied by the factor $<+|V_3|+><+|V_4|+>$. Note that ψ_2^* means conjugate transpose of ψ_2 which is therefore a $1 \times N^3$ row vector. $I_{N^3} \otimes \psi_2^*$ and $\psi_2^* \otimes I_{N^3}$ are thus $N^3 \times N^6$ matrices and since V_{12} is a $N^6 \times N^6$ matrix and $|\psi_1 \otimes \psi_2>$ is an $N^6 \times 1$ vector, the quantities $(I_{N^3} \otimes \psi_2^*)V_{12}|\psi_1 \otimes \psi_2>$ and $(\psi_2^* \otimes I_{N^3})V_{12}|\psi_2 \otimes \psi_1>$ are $N^3 \times 1$ column vectors.

Now suppose we have an equation of the form

$$id\psi_1/dt = F_1(\psi_1, \psi_2), id\psi_2/dt = F_2(\psi_1, \psi_2)$$

as in the above derived Hartree-Fock equations. Here $F_k(\psi_1, \psi_2), k = 1, 2$ are $N^3 \times 1$ vector valued functions of $\psi_1, \psi_2 \in \mathbb{C}^{N^3 \times 1}$ and $\psi_1^*, \psi_2^* \in \mathbb{C}^{1 \times N^3}$. These are solved in MATLAB using a for loop implementation: Represent ψ_1, ψ_2 as $N^3 \times K$ matrices where $\psi_1(:, t), \psi_2(:, t)$ are the vectors ψ_1, ψ_2 at time t. K is the total number of time samples. Then the for loop iteration will read

for $t = 1 : K$
$\psi_1(:, t+1) = \psi_1(:, t) - i * \delta * F_1(\psi_1(:, t), \psi_2(:, t));$
$\psi_2(:, t+1) = \psi_2(:, t) - i * \delta * F_2(\psi_1(:, t), \psi_2(:, t));$
$\psi_1(: t+1) = \psi_1(:, t+1)/norm(\psi_1(:, t+1));$
$\psi_2(:, t+1) = \psi_2(:, t+1)/norm(\psi_2(:, t+1));$

[18] Computing the perturbation in the singular values and vectors under a small perturbation of a matrix with application to the MUSIC and ESPRIT algorithms.

$$A \in \mathbb{C}^{M \times N}$$

Has an SVD

$$A = UDV^*$$

Let $rank(A) = r$. Then since

$$A^*A = VDD^TV^*$$

With DD^T diagonal square matrix having exactly r nonzero (positive) elements, say $\lambda_k = \sigma_k^2, k = 1, 2, \ldots, r$, we can write

$$A^*Av_k = \lambda_k v_k, k = 1, 2, \ldots, r, Av_k = 0, k = r+1, \ldots, N$$

Also,

$$AV = UD$$

Gives for the columns $u_1, \ldots, u_M$ of U,

$$u_k = Av_k/\sigma_k, k = 1, 2, \ldots, r$$

Suppose now that A gets perturbed to $A + \delta A$. Then, $B = A^*A$ will get perturbed to $B + \delta B$, where

$$\delta B = A^*.\delta A + \delta A^*.A + \delta A^*.\delta A$$

$$= \delta_1 B + \delta_2 B$$

Where

$$\delta_1 B = A^* \delta A + \delta A^* A$$

is the first order perturbation in B and

$$\delta_2 B = \delta A^* . \delta A$$

Is the second order perturbation in B. Now using second order perturbation theory, we write for the perturbation in v_k, Given by $\delta_1 v_k + \delta_2 v_k$ where $\delta_1 v_k$ and $\delta_2 v_k$ are the first and second order perturbation terms, and also for the corresponding perturbations $\delta_1 \lambda_k + \delta_2 \lambda_k$ in the singular values,

$$(B + \delta_1 B + \delta_2 B)(v_k + \delta_1 v_k + \delta_2 v_k) = (\lambda_k + \delta_1 \lambda_k + \delta_2 \lambda_k)(v_k + \delta_1 v_k + \delta_2 v_k)$$

Equating terms of the same order on both sides gives

$$B v_k = \lambda_k v_k,$$

$$B \delta_1 v_k + \delta_1 B v_k = \lambda_k \delta_1 v_k + \delta_1 \lambda_k . v_k,$$

$$B \delta_2 v_k + \delta_1 \lambda_k \delta_1 v_k + \delta_2 \lambda_k v_k = \lambda_k \delta_2 v_k$$

Perturbation theoretic analysis of the SVD based ESPRIT algorithm

Step 1: The noiseless array signal model is

$$\mathbf{X}_0 = \mathbf{AS}, \mathbf{X}_1 = \mathbf{A\Phi S}$$

and the noisy array signal model is

$$\mathbf{Y}_0 = \mathbf{X}_0 + \mathbf{W}_0, \mathbf{Y}_1 = \mathbf{X}_1 + \mathbf{W}_1$$

Here, $\mathbf{A}$ is an $n \times p$ matrix of full column rank p while $\mathbf{S}$ is a $p \times m$ matrix of full row rank p. Φ is a $p \times p$ diagonal unitary matrix. We also assume that $n \leq m$ and then Y_0, Y_1 are $n \times m$ matrices of full row rank n (This assumption means that the number of time samples is much larger than the number of sensors). We write the SVD's of X_0, Y_0, X_1, Y_1 as

$$X_0 - U_0 D_0 V_0^*, X_1 = U_1 D_1 V_1^*,$$

$$Y_0 = \tilde{U}_0 \tilde{D}_0 \tilde{V}_0^*, Y_1 = \tilde{U}_1 \tilde{D}_1 \tilde{V}_1^*$$

where U_0, U_1 are $n \times p$ matrices each with orthonormal columns, V_0, V_1 are $m \times p$ matrices each with orthonormal columns, D_0, D_1 are $p \times p$ diagonal matrices with positive diagonal entries, $\tilde{U}_0, \tilde{U}_1$ are $n \times n$ unitary matrices, $\tilde{V}_0, \tilde{V}_1$ are $m \times n$ matrices each with orthonormal columns and $\tilde{D}_0, \tilde{D}_1$ are $n \times n$ diagonal matrices with positive diagonal entries. The first p diagonal entries of $\tilde{D}_0$ and $\tilde{D}_1$ are small perturbations of the diagonal entries of D_0 and D_1 respectively. The last $n-p$ diagonal entries of $\tilde{D}_0$ and $\tilde{D}_1$ are small positive perturbations of zero. We can thus write

$$\tilde{D}_0 = diag[D_0 + \delta D_0, \delta D_{01}]$$

$$\tilde{D}_1 = diag[D_1 + \delta D_1, \delta D_{11}]$$

$$\tilde{U}_0 = [U_0 + \delta U_0 | U_{01} + \delta U_{01}],$$

$$\tilde{V}_0 = [V_0 + \delta V_0 | V_{01}]$$

$$\tilde{V}_1 = [V_1 + \delta V_1 | V_{11}],$$

$$\tilde{U}_1 = [U_1 + \delta U_1 | U_{11} + \delta U_{11}]$$

where $\delta D_0, \delta D_{01}, \delta D_1, \delta D_{11}, \delta V_0, V_{01}, \delta V_1, V_{11}$ are computed by standard perturbation theory for Hermitian matrices with the Hermitian matrices $Y_0^* Y_0$ and $Y_1^* Y_1$ being regarded as small perturbations of the Hermitian matrices $X_0^* X_0$ and $X_1^* X_1$ respectively. Note that $X_k^* X_k, k = 0, 1$ are $m \times m$ positive semidefinite matrices of rank p while $Y_k^* Y_k, k = 0, 1$ are $m \times m$ positive semidefinite matrices of rank n. Note that δV_0 and δD_0 and likewise, δV_1 and δD_1 are computed using standard non-degenerate perturbation theory for Hermitian matrices while $V_{01}, \delta D_{01}, V_{11}$ and δD_{11} are computed using degenerate perturbation theory for Hermitian matrices based on the secular matrix theory as described in Dirac's book, "The principles of quantum mechanics". Then,

$$U_0 D_0 = X_0 V_0, U_0 = X_0 V_0 D_0^{-1},$$

$$U_1 D_1 = X_1 V_1, U_1 = X_1 V_1 D_1^{-1}$$

Thus, since $\delta X_0 = Y_0 - X_0 = W_0, \delta X_1 = Y_1 - X_1 = W_1$, we get

$$\delta U_0 = W_0 V_0 D_0^{-1} + X_0 \delta V_0 D_0^{-1} - X_0 V_0 D_0^{-2} \delta D_0$$

$$\delta U_1 = W_1 V_1 D_1^{-1} + X_1 \delta V_1 . D_1^{-1} - X_1 V_1 D_1^{-2} . \delta D_1$$

using first order perturbation theory. The diagonal entries of Φ are derived as the rank reducing numbers (rrn) of the matrix pencil

$$P_1(\gamma) = X_1 - \gamma . X_0 = A(\Phi - \gamma . I_p) S$$

Now,

$$X_1 - \gamma X_0 = U_1 D_1 V_1^* - \gamma . U_0 D_0 V_0^*$$
$$= U_1 (D_1 - \gamma Q_1 D_0 Q_2^*) V_1^*$$

and since U_1 and V_1 have full column rank p, the rrn's of the matrix pencil $P_1(\gamma)$ can alternatively be obtained as the rrn's of the pencil

$$D_1 - \gamma Q_1 D_0 Q_2^*$$

where

$$Q_1 = U_1^* U_0, Q_2 = V_1^* V_0$$

are $p \times p$ matrices. These rrn's are therefore equivalently the solutions of the pth-degree polynomial equation

$$det(D_1 - \gamma Q_1 D_0 Q_2^*) = 0$$

Now let γ_k be one of these rrn's, ie, it is one of the diagonal entries of Φ. Then there exist exactly two vectors ξ_k and η_k upto a constant of proportionality such that

$$\xi_k^*(D_1 - \gamma_k Q_1 D_0 Q_2^*) = 0, (D_1 - \gamma_k Q_1 D_0 Q_2^*)\eta_k = 0$$

These equations can be expressed as

$$\xi_k^* U_1^*(U_1 D_1 V_1^* - \gamma_k U_0 D_0 V_0^*)V_1 = 0,$$

$$U_1^*(U_1 D_1 V_1^* - \gamma_k U_0 D_0 V_0^*)V_1 \eta_k = 0$$

since

$$U_1^* U_1 = V_1^* V_1 = I_p$$

or equivalently,

$$x_k^* A(\Phi - \gamma_k.I_p)SV_1 = 0, U_1^* A(\Phi - \gamma_k I_p)Sy_k = 0 - - - (1)$$

where

$$x_k = U_1 \xi_k, y_k = V_1 \eta_k$$

Now since

$$R(A) = R(U_0) = R(U_1), R(S^*) = R(V_0) - R(V_1)$$

and the second implies

$$N(S) = R(V_1)^\perp$$

while the first implies

$$R(U_1)^\perp = N(A^*),$$

it follows that (1) implies and is in fact equivalent to

$$x_k^* A(\Phi - \gamma_k I_p)S = 0 - - - (2a)$$

and

$$A(\Phi - \gamma_k I_p)Sy_k = 0 - - - (2b)$$

or equivalently, since S has full row rank and A has full column rank,

$$x_k^* A(\Phi - \gamma_k I_p) = 0, (\Phi - \gamma_k I_p)Sy_k = 0$$

Moreover,

$$x_k \in R(U_1) = R(A) = N(A^*)^\perp, y_k \in R(V_1) = R(S^*) = N(S)^\perp$$

it follows that

$$x_k^* A \neq 0, Sy_k \neq 0$$

and hence, (2) when combined with the fact that only the kth diagonal entry of $\Phi - \gamma_k I_p$ vanishes that

$$x_k^* a_k \neq 0, x_k^* a_j = 0, j \neq k,$$

$$s_k^T y_k \neq 0, s_j^T y_k = 0, j \neq k$$

where $s_k^T, k = 1, 2, ..., p$ are the rows of S. Thus,

$$x_k^* X_0 = \xi_k^* AS = x_k^* a_k s_k^T,$$

$$X_0 y_k = ASy_k = a_k s_k^T y_k$$

Note that $a_k, k = 1, 2, ..., p$ are the columns of A while $s_k^T, k = 1, 2, ..., p$ are the rows of S. This gives us the identity that s_k^T is proportional to $x_k^* X_0$ and a_k is proportional to $X_0 y_k$. Note further that the above identities imply

$$(x_k^* a_k)(s_k^T y_k) = x_k^* X_0 y_k$$

Now when noise perturbs the data matrices, we have to consider the matrix pencil

$$\tilde{P}(\gamma) = Y_1 - \gamma.Y_0$$

in place of its unperturbed version $P(\gamma)$. This is a matrix of size $n \times m$ and will generally have rank n but when γ assumes one of its rrn values, the rank of this matrix pencil drops to $n-1$. This is in contrast to the noiseless case where where the rank of $P(\gamma) = X_1 - \gamma.X_0$ (which is again a matrix of size $n \times m$) drops from p to $p-1$ when γ assumes one of the rrn values. Now,

$$\tilde{P}(\gamma) = \tilde{U}_1 \tilde{D}_1 \tilde{V}_1^* - \gamma.\tilde{U}_0 \tilde{D}_0 \tilde{V}_0^*$$

$$= \tilde{U}_1(\tilde{D}_1 - \gamma.\tilde{Q}_1 \tilde{D}_0.\tilde{Q}_2^*)\tilde{V}_1^*$$

where

$$\tilde{Q}_1 = \tilde{U}_1^* \tilde{U}_0, \tilde{Q}_2 = \tilde{V}_1^* \tilde{V}_0$$

$\tilde{D}_0$ is an $n \times n$ diagonal matrix such that its first p diagonal entries are large, these being the corresponding perturbations of those of D_0 while its last $n-p$ diagonal entries are small, these being perturbations of the zero singular values of X_0. We note that since $\tilde{U}_1$ and $\tilde{V}_1$ have full column ranks n, it follows that the rrn's of $\tilde{P}(\gamma)$ are same as those of the $n \times n$ matrix

$$\tilde{D}_1 - \gamma.\tilde{Q}_1.\tilde{D}_0.\tilde{Q}_2^*$$

Note that

$$\tilde{U}_0 = Y_0 \tilde{V}_0.\tilde{D}_0^{-1}, \tilde{U}_1 = Y_1 \tilde{V}_1.\tilde{D}_1^{-1}$$

Note also that $\tilde{Q}_1, \tilde{Q}_2$ are $n \times n$ non-singular matrices. These rrn's are obtained by solving

$$det(\tilde{D}_1 - \gamma.\tilde{Q}_1.\tilde{D}_0.\tilde{Q}_2^*) = 0 - - - (3)$$

p of these solutions will generally be close to the diagonal elements of Φ (ie, the rrn's for the unperturbed case) while the remaining $n-p$ solutions will be close to zero. Now, we arrange our SVD so that the first p largest diagonal values of $\tilde{D}_0$ appear before the remaining $n-p$ diagonal values. Then, we have as discussed above,

$$\tilde{D}_0 = diag[D_0 + \delta D_0|\delta D_{01}]$$

and we are interested in evaluating not the solutions of (3), but rather the solutions of the determinantal equation obtained by considering the top $p \times p$ left hand corner block of

$$\tilde{D}_1 - \gamma.\tilde{Q}_1.\tilde{D}_0.\tilde{Q}_2^*$$

The top left hand corner block of $\tilde{D}_1$ is $D_1 + \delta D_1$ while the top left hand corner block of $\tilde{Q}_1.\tilde{D}_0.\tilde{Q}_2^*$ is

$$(\tilde{Q}_1.\tilde{D}_0.\tilde{Q}_2^*)_{11} = (\tilde{Q}_1)_{11}(D_0 + \delta D_0)(\tilde{Q}_2)_{11})^* + (\tilde{Q}_1)_{12}\delta D_{01}((\tilde{Q}_2)_{12})^*$$

where we have used the partition

$$\tilde{Q}_k = \begin{pmatrix} (Q_k)_{11} & (Q_k)_{12} \\ (Q_k)_{21} & (Q_k)_{22} \end{pmatrix}, k = 1, 2$$

Next observe that using first order perturbation theory,

$$\tilde{Q}_1 = \tilde{U}_1^* \tilde{U}_0 =$$

$$[U_1 + \delta U_1|U_{11} + \delta U_{11}]^*[U_0 + \delta U_0|U_{01} + \delta U_{01}]$$

implies that

$$(\tilde{Q}_1)_{11} =$$

$$(U_1 + \delta U_1)^*(U_0 + \delta U_0) = U_1^* U_0 + U_1^* \delta U_0 + \delta U_1^* U_0$$

$$= Q_1 + U_1^* \delta U_0 + \delta U_1^* U_0$$

$$(\tilde{Q}_1)_{12} = (U_1 + \delta U_1)^*(U_{01} + \delta U_{01}) = U_1^* U_{01} + U_1^* \delta U_{01} + \delta U_1^* U_{01} = U_1^* \delta U_{01} + \delta U_1^* U_{01}$$

$$\tilde{Q}_2 = \tilde{V}_1^* \tilde{V}_0 =$$

$$[V_1 + \delta V_1|V_{11}]^*[V_0 + \delta V_0|V_{01}]$$

implies that

$$(\tilde{Q}_2)_{11} = V_1^* V_0 + V_1^* \delta V_0 + \delta V_1^* V_0 =$$

$$Q_2 + V_1^* \delta V_0 + \delta V_1^* V_0$$

$$(\tilde{Q}_2)_{12} = V_1^* V_{01} + \delta V_1^* V_{01} = \delta V_1^* V_{01}$$

and, thus,

$$(\tilde{Q}_1.\tilde{D}_0.\tilde{Q}_2^*)_{11} =$$

$$Q_1 D_0 Q_2 + Q_1 \delta D_0.Q_2^*$$

upto first order perturbation terms. Thus upto first order perturbation theory, the estimated values $\hat{\gamma}_k, k = 1, 2, ..., p$ of the true rrn's $\gamma_k, k = 1, 2, ..., p$ are the solutions of

$$det(D_1 + \delta D_1 - \hat{\gamma}.Q_1(D_0 + \delta D_0)Q_2^*) = 0$$

and associated left and right eigenvectors corresponding to the estimated rrn's $\hat{\gamma}_k$ are obtained by solving

$$\hat{\xi}_k^T(D_1 + \delta D_1 - \hat{\gamma}_k.Q_1(D_0 + \delta D_0)Q_2^*) = 0,$$

$$(D_1 + \delta D_1 - \hat{\gamma}.Q_1(D_0 + \delta D_0)Q_2^*)\hat{\eta}_k = 0$$

Remarks:

[a]

$$[V_0|V_{01}]$$

is a matrix of size $m \times n$ having orthonormal columns. V_{01} is determined by the eigenvectors of the "secular perturbation matrix" corresponding to the zero (unperturbed) eigenvalue of the $m \times m$ matrix $X_0^*X_0$ and its perturbation

$$\delta(X_0^*X_0) = X_0^*W_0 + W_0^*X_0$$

The secular matrix corresponding to this is the $m \times m$ matrix of this perturbing operator w.r.t an onb for the zero eigenvalue subspace of $X_0^*X_0$. This secular matrix is therefore of size $m - p \times m - p$. The $n - p$ columns of V_{01} are thus orthonormal and form a subspace of the space spanned by the $m-p$ eigenvectors of $X_0^*X_0$ corresponding to its zero eigenvalue. Likewise, the $n - p$ columns of V_{11} are also orthonormal and form a subspace of the space spanned by the $m-p$ eigenvectors of $X_0^*X_0$ corresponding to its zero eigenvalue. This latter $m - p$ dimensional space is precisely $N(X_0^*X_0) = N(X_0) = N(S) = N(X_1^*X_1) = N(X_1)$. Equivalently, $R(V_0) = R(V_1) = R(S^*)$ and $N(S) = R(S^*)^\perp$ contains $R(V_{01})$ as well as $R(V_{11})$. $R([V_0 + \delta V_0|V_{01}]) = R(Y_0^*) = R(Y_0^*Y_0)$ and$R([V_1 + \delta V_1|V_{11}]) = R(Y_1^*) = R(Y_1^*Y_1)$ and these two subspaces clearly have dimension n. $R(V_0) = R(V_1) = R(S^*) = N(S)^\perp$ is in particular orthogonal to $R(V_{01})$ as well to $R(V_{11})$.In particular, $V_1^*V_{01} = 0$

[b] Similar remarks apply to U in place of V. The corresponding relations are obtained using $X_0^* = S^*A^*, X_1^* = S^*\Phi^*A^*, Y_0^* = X_0^* + W_0^*, Y_1^* = X_1^* + W_1^*$ in place of X_0, X_1, Y_0, Y_1. In particular, $U_1^*U_{01} = 0$. This could also be seen directly using

$$[U_0 + \delta U_0|U_{01}]diag[D_0 + \delta D_0, \delta D_{01}] = (X_0 + W_0)[V_0 + \delta V_0|V_{01}]$$

which implies using first order perturbation theory that

$$U_0D_0 = X_0V_0,$$

$$U_0\delta D_0 + \delta U_0D_0 = X_0\delta V_0 + W_0V_0,$$

$$X_0 V_{01} = 0$$

$$U_{01}\delta D_{01} = W_0 V_{01}$$

Thus in particular,

$$U_{01} = W_0 V_{01}(\delta D_{01})^{-1}$$

These equations do not appear to imply $U_1^* U_{01} = 0$. However, let us see the contribution of this term to

$$(\tilde{Q}_1)_{12}\delta D_{01}((\tilde{Q}_2)_{12})^*$$

It is given by

$$U_1^* U_{01}\delta D_{01}((\tilde{Q}_2)_{12})^* =$$

$$U_1^* W_0 V_{01}((\tilde{Q}_2)_{12})^*$$

$$= U_1^* W_0 V_{01}\delta V_1^* V_{01}$$

which is of the second order of smallness in perturbation theory and hence can be neglected. In fact, this could directly be inferred from the fact that $(\tilde{Q}_1)_{12}\delta D_{01}((\tilde{Q}_2)_{12})^*$ is of the second order of smallness since $((\tilde{Q}_2)_{12})^* = \delta V_1^* V_{01}$ is of the second order of smallness.

[19] Problem on video-conferencing (Suggested to me by Prof.Vijyant Agarwal). There are N speakers numbered $1, 2, ..., N$ conversing over a common line. Let $\mathbf{x}_k$ denote the speech vector signal spoken by the kth speaker. Assume that the listener receives a superposition of compressed versions of the different speakers. For example, if $x_k(t), t = 1, 2, ..., M$ are the speech samples of the kth speaker, then his dominant wavelet coefficients are

$$c_k(n,m) = \sum_{t=1}^{N} x_k(t)\psi_{nm}(t), (n,m) \in D_k$$

where D_k is a small index pair set compared with the original number N of time samples of the signal. This equation can be expressed in the form

$$\mathbf{c}_k = \mathbf{A}_k \mathbf{x}_k$$

where

$$\mathbf{A}_k = ((\psi_{nm}(t)))_{(n,m)\in D_k, t\in\{1,2,...,N\}}$$

Let $\mu(D_k)$ denote the number of elements in D_k. Then

$$\mathbf{A}_k \in \mathbb{R}^{\mu(D_k)\times N}$$

and since we are assuming that

$$\mu(D_k) << N$$

it means that $\mathbf{A}_k$ has very few rows compared to the number of its columns. Thus $\mathbf{c}_k \in \mathbb{R}^{\mu(D_k)}$ is a wavelet compressed version of $\mathbf{x}_k \in \mathbb{R}^N$. The listener receives the superposition

$$\mathbf{c} = \sum_{k=1}^{N} \alpha(k)\mathbf{c}_k = \sum_{k=1}^{N} \alpha(k)\mathbf{A}_k\mathbf{x}_k$$

The listener also has some idea of what the original speech $\mathbf{x}_k$ of the kth speaker is for each k. This approximate signal is $\mathbf{w}_k$. He also wishes his estimate of each speaker's speech to have a small energy. Thus, taking all these considerations into account, the listener must minimize

$$E(\mathbf{x}_k, k=1,2,...,N,\lambda) = \sum_{k=1}^{N} \beta(k) \parallel \mathbf{x}_k - \mathbf{w}_k \parallel^2 + \sum_{k=1}^{N} \gamma(k) \parallel \mathbf{x}_k \parallel^2 + \lambda(\mathbf{c} - \sum_{k=1}^{N} \alpha(k)\mathbf{A}_k\mathbf{x}_k)$$

where λ is a Lagrange multiplier used to incorporate the constraint of the compressed superposition of the speech signals spoken by the different speakers. E must first be minimized w.r.t $\mathbf{x}_k, k = 1,2,...,N,\lambda$, then a performance analysis must be carried out along the following lines:

[a] Assume that $\mathbf{w}_k = \mathbf{x}_k + \delta\mathbf{x}_k, k = 1,2,...,N$ where the $\delta\mathbf{x}_k^{\prime}s$ are small, say these are bounded in norm by ϵ_k:

$$\parallel \delta\mathbf{x}_k \parallel \leq \epsilon_k, k = 1,2,...,N$$

Then derive upper bounds on

$$\parallel \mathbf{x}_k - \hat{\mathbf{x}}_k \parallel$$

[b] Assume that the the $\alpha(k), \beta(k), \gamma(k)$ are not precisely known due to system uncertainties, ie, although we use $\{\alpha(k), \beta(k), \gamma(k)\}$ during the minimization process, while implementing, the estimator, we use $\alpha(k) + \delta\alpha(k), \beta(k) + \delta\beta(k), \gamma(k) + \delta\gamma(k)$. Then how will the values of $\parallel \mathbf{x}_k - \hat{\mathbf{x}}_k \parallel$ change ?

Setting the partial derivatives of E w.r.t $\mathbf{x}_k$ at the estimate $\hat{\mathbf{x}}_k$ and w.r.t. λ to zero then gives

$$2\beta(k)(\hat{\mathbf{x}}_k - \mathbf{w}_k) + 2\gamma(k)\hat{\mathbf{x}}_k - \alpha(k)\mathbf{A}_k^T\lambda = 0,$$

$$\mathbf{c} = \sum_{k} \alpha(k)\mathbf{A}_k\hat{\mathbf{x}}_k$$

Thus,

$$2(\beta(k) + \gamma(k))\hat{\mathbf{x}}_k = 2\beta(k)\mathbf{w}_k + \alpha(k)\mathbf{A}_k^T\lambda$$

$$\hat{\mathbf{x}}_k = (\frac{\beta(k)}{\beta(k) + \gamma(k)})\mathbf{w}_k + (\frac{\alpha(k)}{2(\beta(k) + \gamma(k))})\mathbf{A}_k^T\lambda$$

and then

$$\mathbf{c} = \sum_k (\frac{\alpha(k)\beta(k)}{\beta(k)+\gamma(k)})\mathbf{A}_k\mathbf{w}_k + \sum_k (\frac{\alpha(k)^2}{2(\beta(k)+\gamma(k))})\mathbf{A}_k\mathbf{A}_k^T\lambda$$

and therefore,

$$\lambda = [\sum_k (\frac{\alpha(k)^2}{2(\beta(k)+\gamma(k))})\mathbf{A}_k\mathbf{A}_k^T]^{-1}.[\mathbf{c} - \sum_k \frac{\alpha(k)\beta(k)}{\beta(k)+\gamma(k)}\mathbf{A}_k\mathbf{w}_k]$$

and thus,

$$\hat{\mathbf{x}}_k =$$

$$(\frac{\beta(k)}{\beta(k)+\gamma(k)})\mathbf{w}_k + (\frac{\alpha(k)}{2(\beta(k)+\gamma(k))})\mathbf{A}_k^T[\sum_m (\frac{\alpha(m)^2}{2(\beta(m)+\gamma(m))})\mathbf{A}_m\mathbf{A}_m^T]^{-1}.[\mathbf{c}$$

$$-\sum_m \frac{\alpha(m)\beta(m)}{\beta(m)+\gamma(m)}\mathbf{A}_m\mathbf{w}_m]$$

$$= b(k)\mathbf{w}_k + a(k)\mathbf{A}_k^T.[\sum_m e(m)\mathbf{A}_m\mathbf{A}_m^T]^{-1}.[\mathbf{c} - \sum_m d(m)\mathbf{A}_m\mathbf{w}_m]$$

where

$$a(k) = \frac{\alpha(k)}{2(\beta(k)+\gamma(k))}, b(k) = \frac{\beta(k)}{\beta(k)+\gamma(k)},$$

$$d(k) = \frac{\alpha(k)\beta(k)}{\beta(k)+\gamma(k)}, e(k) = \frac{\alpha(k)^2}{2(\beta(k)+\gamma(k))}$$

Statistical performance analysis: Let

$$\mathbf{w}_k = \mathbf{x}_k + \delta\mathbf{x}_k$$

where $\delta\mathbf{x}_k, k = 1, 2, ..., N$ are random vectors assumed to be small random perturbations of the true speech signal vectors of the speakers. We also assume that $\mathbf{c} = \sum_k \alpha(k)\mathbf{A}_k\mathbf{x}_k$ is known to the listener. We then obtain using the above formula

[20] Application of supersymmetry to the design of quantum unitary gates.

[a] What is meant by a supersymmetric theory of elementary particles ?

Syllabus for End-Semester Examination

EC-SPC03

[1] Axioms of probability theory, probability spaces, random variables, expectation, properties of the expectation map, Conditional expectation w.r.t a sub-sigma algebra derivation using Radon-Nikodym derivatives and another derivation using orthogonal projections in Hilbert space and density of $L^2(P)$ in $L^1(P)$.

[2] Chebyshev's inequality, Borel-Cantelli lemmas, definition of convergence of a sequence of random variables in distribution, in probability, in mean square, almost surely (ie, with probability 1). Proof that convergence in mean square

implies convergence in probability, convergence almost surely implies convergence in probability implies convergence in distribution. L^p convergence and Holder's inequality.

[3] Filtration on a probability space, martingales, submartingale and supermartinagles w.r.t a filtration, Doob's maximal inequality for martingales, the martingale upcrossing and downcrossing inequality for martingales, Kolmogorov's inequality for sums of independent random variables as a special case of Doob's maximal inequality, Doob's sub-martingale convergence theorem using upcrossing inequality for submartingales.

[4] Weak and strong laws of large numbers. Proof of the weak law of large numbers using Chebyshev's inequality, proof of the strong law using Kolmogorov's inequality for sums of independent random variables.

[5a] Bayesian binary and multiple hypothesis testing, Neyman-Pearson test, the likelihood ratio test, binary hypothesis testing for sequences of iid random variables, the asymptotic rate at which the optimal probability of miss converges to zero as the relative entropy between the two probability distribution given that the false alarm probability converges to zero (Stein's theorem) (proof based on the large deviation principle)

[5b] Parameter estimation based on measured data using Maximum likelihood, Maximum aposteriori (MAP) and Minimum mean square error (MMSE) criteria. Asymptotic properties of the maximum likelihood estimator when the measurement is a sequence of iid random variables. The Cramer-Rao lower bound on the variance of unbiased and biased parameter estimators.

[6] Stationarity and ergodicity of a discrete time stochastic process. Conditions for mean and correlation ergodicity of a stochastic process with application to Gaussian processes. Application of the ergodic theorem to estimation theory.

[7a] Estimation of parameters in linear models, linear prediction theory for wide sense stationary processes. The optimum causal Wiener filter derivation using Wiener's spectral factorization method and using Kolmogorov's innovation process theory.

[7b] Statistical performance analysis of parameter estimation in linear models based on data matrix perturbation theory.

[8] The Levinson-Durbin algorithm for order-recursive linear prediction for WSS processes with known statistics. Derivation based on Toeplitz and centrosymmetric properties of correlation matrices. Forward and backward prediction filter update formulas. Block diagrammatic representation of lattice filter.

Order recursive construction of joint process filter using the backward prediction filters and orthogonality of the backward prediction errors of different orders.

[9] The projection operator update formula. Another derivation of the Levinson-Durbin algorithm based on this formula.

[10] The RLS algorithm for time recursive prediction and filtering for time series having unknown statistical correlations.

[11] The RLS-Lattice algorithm for joint time and order recursive prediction.

[12] The Levinson-Durbin, RLS and RLS lattice filters for multivariate time series.

[13] Brownian motion and white Gaussian noise. Einstein's derivation of the diffusion equation for the pdf of Brownian motion, Einstein's evaluation of the diffusion constant of Brownian motion in terms of viscosity, temperature, Boltzmann's constant and the radius of the pollen particle executing Brownian motion. White noise as the formal derivative of Brownian motion, correlation properties of Brownian motion and white noise, Ito's formula for Brownian motion and the Levy oscillation property.

[14] Independent increment processes, Brownian motion and Poisson processes as special case. The characteristic functional of a superposition of Brownian motion and independent Poisson processes.

[15] Adapted processes and stoptimes w.r.t to a filtration. Doob's optional sampling theorem for martingales, exponential martingales for Brownian motion, application of the optional sampling theorem for martingales to computing the statistics of the first hitting time of Brownian motion at a given level.

[16] Construction of the Ito stochastic integral w.r.t Brownian motion for L^2-adapted processes. Properties or the Ito stochastic integral (The Ito stochastic integral as an isomoprhism between Hilbert spaces).

[17] The Ito stochastic differential equation, proof of existence and uniqueness of the solutions using the Lipschitz conditions on the drift and diffusion coefficients and Doob's martingale inequality, derivation of the forward and Backward Fokker-Planck-Kolmogorov equations for the transition probabilty density of a diffusion process using Ito's formula.

[18] The periodogram power spectral density estimator for a stationary discrete time Gaussian process. Properties of asymptotic mean and covariance of the periodogram. Inconsistency of the periodogram. Improvement on the variance of the periodogram estimator using smoothening windows applied to the data.

[19] MUSIC and ESPRIT as high resolution eigensubspace based estimators of frequencies and directions of arrival for plane wave sources incident upon an array of sensors. Finite data performance analysis of the MUSIC and ESPRIT estimators based on matrix perturbation theory, SVD based MUSIC and ESPRIT algorithm using data matrices, performance analysis of SVD based algorithms.

[20] Kushner-Kallianpur filter, Extended Kalman filter and Belavkin quantum filter for estimating the state of a process with noisy measurements.

[21] Application of the Belavkin filter to estimating the spin of the electron interacting with a magnetic field and subject to quantum noise based on the Hudson-Parthasarathy-noisy Schrodinger equation.

[22] The energy of an electromagnetic field within a cavity resonator, quantization of this energy using field creation and annihilation operators, cavity field interacting with the bath noisy field, Application of the Belavkin filter for estimating the state of the cavity field.

[23] The cavity electromagnetic field along with the cavity Dirac electron-positron field in interacting with the bath field. Application of the Belavkin filter to estimating the cavity electromagnetic field and the cavity electron-positron field from non-demolition bath measurements.

[24] Description of the bath electron-positron field based on Fermionic creation and annihilation processes.

[25] Solving Dirac's relativistic wave equation for an electron in a radial potential. Perturbation of the time dependent Dirac equation by electromagnetic quantum noise.

[26] Group theoretic statistical image processing.

[a] Properties of the 3-dimensional rotation group and its Lie algebra.

[b] Properties of the d-dimensional Euclidean motion group (rotations and translations of $\mathbb{R}^3$) and its Lie algebra.

[c] Computing the Haar measure on the rotation group.

[d] Irreducible representations of a compact group and the Schur orthogonality relations, character of a representation. Representations of a Lie algebra and its relation to the representation of the corresponding group.

[e] The Peter-Weyl theorem for compact groups. Proof based on the spectral theory of compact Hermitian operators in a Hilbert space.

[f] The irreducible representations of the 3-D rotation group in terms of spherical harmonic functions. Proof of the construction based on properties of the angular momentum operators.

[g] Application of the group representation theoretic Fourier transform to estimate the scale, rotation and translation of a 3-D image field in the presence of blurring noise.

[h] Statistical performance analysis of the rotation estimation algorithm.

[i] The Frobenius-Mackey theory of induced representations of a semidirect product applied to the construction of the irreducible representations of the 3-D Euclidean motion group of rotations and translations.

[l] Determining invariants of a group action on image fields using characters with application to pattern classification/invariant feature extraction.

[m] The irreducible representations of $SL(2,\mathbb{C})$ with application to determining the irreducible representations of the Lorentz group:Principal series and supplimentary series of Gelfand.

[n] The irreducible representations of $SL(2,\mathbb{R})$:Principal series and the discrete series of HarishChandra. Relationship of $SL(2,\mathbb{R})$ to planar Lorentz transformations ie, transformations in the $t-y-z$ space.

[o] The Frobenius-Young theory of irreducible representations of the permutation group. Frobenius character formula of the permutation group in terms of generating functions. Applications of the generating function of Frobenius character formula to determining invariant features of the radiation field produced by a set of point objects under a permutation. Estimating the permutation element from the signal collected from the radiation patterns generated by an array of point sources and by a permutation of this array using permutation group representations.

[27] Short notes on supersymmetric field theories.

[a] What are Majorana Fermionic anticommuting variables $\theta = (\theta^a : a = 1,2,3,4)$?

[b] A superfield $S(x,\theta)$ is an arbitrary smooth function of the four Bosonic space-time coordinates $x = (x^\mu : \mu = 0,1,2,3)$ and the four Fermionic coordinates θ. It can be expressed as a fourth degree polyonmial in the Fermionic variables with coefficients being functions of the Bosonic variables.

[c] What are supersymmetry generators $\{Q_a, \bar{Q}_a : a = 0,1,2,3\}$? They satisfy the anticommutation relations

$$\{Q_a, \bar{Q}_b\} = \gamma^\mu_{ab} P_\mu$$

where P_μ is the Bosonic four momentum. Here

$$\bar{Q} = Q^T\gamma^5\epsilon, \gamma^5 = [I_2, -I_2], \epsilon = diag[i\sigma_y, -i\sigma_y]$$

A realization of supersymmetry generators is provided by super-vector fields

$$L_a = (\gamma^\mu\theta)_a\partial/\partial x^\mu + (\gamma^5\epsilon)_{ab}.\partial/\partial\theta^b$$

and

$$\bar{L}_a = (\gamma^5\epsilon)_{ab}L_b$$

Show using the Fermionic anticommutation analog of the Bosonic commutation relations

$$[\partial/\partial x^\mu, x^\nu] = \delta^\nu_\mu$$

namely,

$$\{\partial/\partial\theta^a, \theta^b\} = \delta^b_a$$

that

$$\{L_a, \bar{L}_b\} = \gamma^\mu_{ab}\partial/\partial x^\mu$$

Note that

$$P_\mu = i\partial/\partial x^\mu$$

and hence the operators $\{L_a, \bar{L}_a\}$ provide us with a representation of the supersymmetric Lie algebra generated by $\{J_{\mu\nu}, P_\mu, Q_a, \bar{Q}_a\}$ where $J_{\mu\nu}$ are the four angular momentum operators

$$J_{\mu\nu} = x_\mu P_\nu - x_\nu P_\mu$$

Note that $\{J_{\mu\nu}, P_\mu\}$ generate the Poincare Lie algebra consisting of Lorentz transformations and space-time translations. So here by incorporating Fermionic operators into this Lie algebra, we obtain the super-Poincare Lie algebra.

Supersymmetric current and its conservation: Let $S(x,\theta)$ be a superfield and let $\chi_m(x), m = 1, 2, ...$ denote its component fields. Let $\mathcal{L}$ be a supersymmetric Lagrangian constructed from these component fields. Under an infinitesimal supersymmetry transformation, $\mathcal{L}$ changes by a total four space-time divergence, ie,

$$\delta\mathcal{L} = \partial_\mu J^\mu$$

We can also associate a Noether current N^μ associated with with the change in the Lagrangian under this infinitesimal supersymmetric transformation of the component fields:

$$N^\mu = \frac{\partial\mathcal{L}}{\partial\chi_{m,\mu}}\delta\chi_m$$

The Noether current is conserved when the field equations are satisfied provided that the Lagrangian is invariant under the supersymetry transformation follows immediately by making use of the Euler-Lagrange equations. We have in fact, from the Euler-Lagrange equations,

$$\partial_\mu \frac{\partial \mathcal{L}}{\partial \chi_{m,\mu}} = \frac{\partial \mathcal{L}}{\partial \chi_m}$$

and therefore,

$$\partial_\mu N^\mu = (\partial \mathcal{L}/\partial \chi_m)\delta\chi_m + \frac{\partial \mathcal{L}}{\partial \chi_{m,\mu}}\delta\chi_{m,\mu} = \delta\mathcal{L}$$

provided that the equations of motion are satisfied. This is zero only when the Lagrangian is invariant under a supersymmetry transformation. However, in general, under a supersymmetry transformation as we noted above, the Lagrangian changes by a four divergence and is not invariant. In other words, only the action integral is supersymmetry invariant. Thus, in general, we can only write

$$\partial_\mu N^\mu = \delta L$$

provided that the equations of motion are satisfied. It follows that the difference of the two equations gives a conservation law:

$$\partial_\mu (S^\mu - N^\mu) = 0$$

in the general case when the equations of motion are satisfied.

Remark: $\partial_\mu J^\mu = \delta L$ is always true while $\partial_\mu N^\mu = \delta L$ is true only when the equations of motion are satisfied. Therefore, in particular, both the equations are valid when the equations of motion are satisfied.

Left and right superderivatives:

$$\theta_L = (1+\gamma^5)\theta/2, \theta_R = (1-\gamma^5)\theta/2$$

Then,

$$D = \gamma^\mu \theta.\partial_\mu - \gamma^5 \epsilon \partial_\theta$$

$$D_L = (1+\gamma^5)\theta/2 = \gamma^\mu \theta_R \partial_\mu - \epsilon \partial_{\theta_L}$$

$$D_R = (1-\gamma^5)\theta/2 = \gamma^\mu \theta_L \partial_\mu + \epsilon \partial_{\theta_R}$$

Note that

$$\gamma^5 \gamma^\mu = -\gamma^\mu \gamma^5$$

Further, it is clear that

$$(1-\gamma^5).D_L = 0, (1+\gamma^5)D_R = 0$$

and hence only two of the $D_L's$ are linearly independent and likewise only two of the $D_R's$ are independent. Further any two of the $D_L's$ anticommute and also any two of the $D_R's$ anticommute:

$$\{D_L, D_L^T\} = \{D_R, D_R^T\} = 0$$

because any two $\theta_R's$ anticommute, any two $\partial_{\theta_L}'s$ anti commute and θ_R anticommutes with ∂_{θ_L} and likewise, θ_L anticommutes with ∂_{θ_R}. Since any two of the $D_R's$ anticommute and since there are only two linearly independent $D_R's$, it follows easily that the product of any three or more $D_R's$ is zero and likewise, the product of any three or more $D_L's$ is also zero.

Further,

$$\{\theta_L, \partial_{\theta_L}^T\} = \{\partial_{\theta_L}, \theta_L^T\} = (‘1 + \gamma^5)/2,$$

and likewise

$$\{\theta_R, \partial_{\theta_R}^T\} = \{\partial_{\theta_R}, \theta_R^T\} = (1 - \gamma^5)/2,$$

It follows that

$$\{D_L, D_R^T\} = \{\gamma^\mu \theta_R \partial_\mu - \epsilon \partial_{\theta_L}, (\gamma^\mu \theta_L \partial_\mu + \epsilon \partial_{\theta_R})^T\}$$
$$= [-\gamma^\mu(1-\gamma^5)\epsilon/2 - \epsilon(1+\gamma^5)\gamma^{\mu T}/2]\partial_\mu$$
$$= [-\gamma^\mu(1-\gamma^5)\epsilon/2 - \gamma^\mu\epsilon(1-\gamma^5)/2]\partial_\mu$$
$$= -\gamma^\mu(1-\gamma^5)\epsilon\partial_\mu$$

This equation is the same as

$$\{D_{La}, D_{Rb}\} = -[\gamma^\mu(1-\gamma^5)\epsilon]_{ab}\partial_\mu$$

which is equivalent to

$$\{D_{Ra}, D_{Lb}\} = -[\gamma^\mu(1-\gamma^5)\epsilon]_{ba}\partial_\mu$$

or equivalently,

$$\{D_R, D_L^T\} = -[\gamma^\mu(1-\gamma^5)\epsilon]^T\partial_\mu$$
$$= \epsilon(1-\gamma^5)\gamma^{\mu T}\partial_\mu$$
$$= \epsilon\gamma^{\mu T}(1+\gamma^5)\partial_\mu = \gamma^\mu\epsilon(1+\gamma^5)\partial_\mu$$

A left Chiral superfield is by definition any function of θ_L and

$$x_+^\mu = x^\mu + \theta_R^T \epsilon \gamma^\mu \theta_L$$

We prove that a superfield Φ is left Chiral iff $D_R\Phi = 0$. The necessity part will follow if we can show that $D_R\theta_L = 0$ and $D_R x_+^\mu = 0$. But

$$D_R\theta_L = [\gamma^\mu\theta_L\partial_\mu + \epsilon\partial_{\theta_R}]\theta_L = 0$$

since

$$\partial_\mu\theta_L = 0 and \partial_{\theta_R}\theta_L = 0$$

Also,

$$D_R x_+^\nu = [\gamma^\mu\theta_L\partial_\mu + \epsilon\partial_{\theta_R}](x^\nu + \theta_R^T\epsilon\gamma^\nu\theta_L)$$

$$= \gamma^\nu \theta_L + \epsilon \partial_{\theta_R} (\theta_R^T \epsilon \gamma^\nu \theta_L$$

$$= \gamma^\nu \theta_L + \epsilon^2 \gamma^\nu \theta_L = 0$$

since

$$\epsilon = diag[e, e], e = i\sigma^2$$

implies

$$\epsilon^2 = -I_4$$

To prove the converse, define

$$x_-^\mu = x^\mu - \theta_R^T \epsilon \gamma^\mu \theta_L$$

Then, is clear that any superfield can be expressed as a function of $\theta_L, \theta_R, x_+^\mu and x_-^\mu$. So it suffices to prove that

$$D_R \theta_R, D_R x_-^\mu$$

are no-zero. This follows at once.

Superfield equations: We know that if $S(x, \theta)$ is a superfield, then under a supersymmetry transformation, $[S]_D$ changes by a four space-time divergence and hence the integral $\int [S]_D d^4x$ is supersymmetry invariant, ie,

$$\int [\bar{\alpha} LS] d^4x = 0$$

where $\bar{\alpha} = \alpha^T \gamma^5 \epsilon$ with α a Majorana Fermionic parameter. Now it is well known that the class of left Chiral superfields is invariant under a supersymmetry transformation and so is the class of right Chiral superfields. To see this, we need only note that

$$L(\theta_L)^T = (\gamma^\mu \theta \partial_\mu + \gamma^5 \epsilon \partial_\theta) \theta_L^T$$

$$= \gamma^5 \epsilon (1 + \gamma^5)/2 = \epsilon (1 + \gamma^5)/2$$

which is a constant matrix and further,

$$L x_+^\nu = (\gamma^\mu \theta \partial_\mu + \gamma^5 \epsilon \partial_\theta)(x^\nu + \theta_R^T \epsilon \gamma^\nu \theta_L)$$

$$= \gamma^\nu \theta + \gamma^5 \epsilon \partial_\theta \theta_R^T \epsilon \gamma^\nu \theta_L$$

$$= \gamma^\nu \theta + \gamma^5 \epsilon ((1 - \gamma^5) \epsilon \gamma^\nu \theta_L + (1 + \gamma^5) \epsilon \gamma^\nu \theta_R)/2$$

$$= \gamma^\nu \theta + \gamma^5 \epsilon . \epsilon \gamma^\nu (\theta_L + \theta_R)$$

$$= 0$$

thus proving the claim. Note that we have used the facts that

$$\theta = \theta_L + \theta_R,$$

$$(\epsilon \gamma^\nu)^T = -\epsilon \gamma^\nu$$

and hence, since θ_L and θ_R anticommute,

$$\theta_R^T \epsilon \gamma^\nu \theta_L = \theta_L^T \epsilon \gamma^\nu \theta_R$$

Let Φ be left invariant Chiral field. We claim that if $[\Phi]_F$ denotes the coefficient of $\theta_L^T \epsilon \theta_L$ in Φ, then $[\Phi]_F$ changes by a four spatio-temporal divergence under a supersymetry transformation. To see this, using the left Chiral property of Φ,we expand Φ as

$$\Phi(\theta_L, x_+^\mu) = \phi_1(x_+) + \theta_L^T \epsilon \phi_2(x_+)$$

$$+\theta_L^T \epsilon \theta_L . \phi_3(x_+)$$

Note that product of three or more of the $\theta_L' s$ is zero and hence the last term is

$$\theta_L^T \epsilon \theta_L . \phi_3(x_+) = \theta_L^T \epsilon . \phi_3(x)$$

Now applying the infinitesimal supersymmetry transformation $\bar{\alpha} L$ to Φ gives

$$\alpha^T \gamma^5 \epsilon . L\Phi$$

and we find that

$$L\Phi = (\gamma^\mu \theta \partial_\mu + \gamma^5 \epsilon . \partial_\theta)\Phi$$

and we easily see that the term in this expression that is quadratic in the $\theta' s$ is given by

$$\gamma^\mu \theta . (\theta_L^T \epsilon \phi_{2,\mu}(x)) +$$

$$+\gamma^5 \epsilon . \partial_\theta (\theta_L^T \epsilon . \phi_{2,\mu}(x) \theta_R^T \epsilon \gamma^\mu \theta_L)$$

and it is easily seen that this is a perfect four space-time divergence and in particular, the coefficient of $\theta_L^T \epsilon \theta_L$ in a perfect space-time four divergence. This proves the claim. Now, if Φ is left Chiral, then for any function f of one variable, $f(\Phi)$ (to be interpreted as a power series in Φ is again left Chiral, since it is also a function of θ_L and x_+^μ only and hence $[f(\Phi)]_F$ is a candidate for a supersymmetric Lagrangian. If $K(x,y)$ is a function of two variables and Φ is left Chiral, then $[K(\Phi^*, \Phi)]_D$ is also a candidate for a supersymmetric Lagrangian. Thus we take for our matter field supersymmetric Lagrangian

$$L_M = [K(\Phi^*, \Phi)]_D + [f(\Phi)]_F$$

In quantum field theory, the matter fields are composed of scalar Klein-Gordon fields, the Dirac electron-positron field and the Yang-Mills matter field the latter is an extended version of the Dirac electron-positron field and includes particles like nucleons.

Remark: The construction of conserved currents and the associated symmetry generated by the corresponding charge is familiar even in classical mechanics. For example, suppose $L(q, q')$ is a Lagrangian which changes by a total time derivative under the infinitesimal symmetry $q \to q + \delta q(q, q')$. Then we can write

$$(\partial L/\partial q)\delta q + (\partial L/\partial q')\delta q' = dF(q, q')/dt$$
$$= (\partial F/\partial q)q' + (\partial F/\partial q')q''$$

Here, we are assuming that F depends only on q, q'. Now,

$$\delta q' = (\partial \delta q/\partial q)q' + (\partial \delta q/\partial q')q''$$

and hence equating the coefficients of q'' on both sides gives

$$(\partial L/\partial q')(\partial \delta q/\partial q') = \partial F/\partial q' - - - (1a)$$

or equivalently,

$$p\partial \delta q/\partial q' = \partial F/\partial q' - - - (1b)$$

and thus, we also have

$$(\partial L/\partial q)\delta q + (\partial L/\partial q')(\partial \delta q/\partial q)q'$$
$$= (\partial F/\partial q)q' - - - (2a)$$

or equivalently,

$$(\partial L/\partial q)\delta q + p(\partial \delta q/\partial q)q'$$
$$= (\partial F/\partial q)q' - - - (2b)$$

When the Euler-Lagrange equations of motion are satisfied, it then follows that

$$d/dt((\partial L/\partial q')\delta q - F) =$$

$$d/dt(\partial L/\partial q')\delta q + (\partial L/\partial q')((\partial \delta q/\partial q)q' + (\partial \delta q/\partial q')q'') - dF/dt$$
$$= (\partial L/\partial q)\delta q + (\partial L/\partial q')((\partial \delta q/\partial q)q' + (\partial \delta q/\partial q')q'') - dF/dt = 0$$

In fact, this conservation law does not depend on using the fact that F is a function of only q, q'. It can be any function of time. However, suppose we introduce the conserved charge

$$Q = (\partial L/\partial q')\delta q - F = p\delta q - F$$

We just showed that when the equations of motion are satisfied,

$$Q' = 0$$

We now assume that $F = F(q, q')$. Then $Q = Q(q, q')$ and therefore Q is an observable in the context of classical mechanics. Further, by virtue of (1),

$$\{Q, q\} = \{p, q\}\delta q + p\{\delta q, q\} - \{F, q\}$$

$$= \delta q + p\partial \delta q/\partial q'\{q', q\} - (\partial F/\partial q')\{q', p\}$$
$$= \delta q + (p\partial \delta q/\partial q' - \partial F/\partial q')\{q', p\}$$

$$= \delta q$$

This equation is true under all conditions, ie, even when the equations of motion are not satisfied. Further, under the conditions that the equations of motion are satisfied,

$$\delta q' = \{Q, q\}' = \{Q', q\} + \{Q, q'\} = \{Q, q'\}$$

Thus, indeed Q generates the symmetry group of transformations that leave the action integral invariant. In the context of fields and supersymmetry, the supesymmetric transformation changes $\mathcal{L}$ by a four space-time divergence:

$$\delta L = \partial_\mu J\mu$$

Here, the role played by F in the above discussion on particle mechanics is played by J^μ and the role played by dF/dt is played by $\partial_\mu J^\mu$. The role played by

$$(\partial L/\partial q)\delta q + (\partial L/\partial q')\delta q'$$

which when the equations of motion are satisfied, equals

$$d/dt((\partial L/\partial q')\delta q)$$

is played by

$$\partial_\mu N^\mu$$

In other words, the role played by

$$p\delta q = (\partial L/\partial q')\delta q$$

is played by N^μ. Thus the role played by the conserved charge $Q = p\delta q - F$ is played by the supersymmetry current $N^\mu - J^\mu$. The role played by the Noether conservation of charge equation

$$Q' = 0$$

is played by the conservation equation of supersymmetry current

$$\partial_\mu(N^\mu - J^\mu) = 0$$

If Φ is a left Chiral superfield, then for any function f of one variable defined as a power series, $f(\Phi)$ is also Left Chiral and hence $[f(\Phi)]_F$ is a supersymmetric Lagrangian and so is $[\Phi^*\Phi]_D$. A candidate Lagrangian for the matter field is then

$$L_M = [\Phi^*\Phi]_D + c_1[f(\Phi)]_F$$

Now, Φ can be expressed as $D_R^2 S$ where S some superfield and $D_R^2 = \bar{D}_R \epsilon D_R$ with

$$\bar{D}_R = \bar{D}(1-\gamma^5)/2 = D^T\gamma^5\epsilon(1-\gamma^5)/2$$
$$= D_R^T\gamma^5\epsilon$$

Recall that since the product of any three $D_R's$ is zero, it follows that $D_R^2 S$ is annihilated by D_R and hence is Left Chiral. Now

$$[f(\Phi)]_F = [f(D_R^2 S)]_F$$

combined with the fact that D_R^2 equals $\partial_{\theta_R}^T \epsilon \partial_{\theta_R}$ plus terms involving the Bosonic derivatives ∂_μ with constant Fermionic parameters implies (using the power series expansion of f that $f(D_R^2 S)$ equals $D_R^2 \tilde{S}$ plus perfect Bosonic divergences for some superfield $\tilde{S}$. Note for example that

$$(D_R^2 S)^2 = D_R^2(S.D_R^2 S) + X$$

where X where X is a perfect Bosonic divergence with constant Fermionic coefficients. Now, $[D_R^2 \tilde{S})]_F$ which is the coefficient of $\theta_L^T \epsilon \theta_L$ in $D_R^2 \tilde{S}$, must coincide with a nonzero real constant c_2 times $[\tilde{S}]_D$ plus a total space-time four divergence. Note that $\theta_L^T \epsilon \theta_L.\theta_R^T \epsilon \theta_R$ equals a non-zero real constant times $(\theta^T \epsilon \theta)^2$. Recall that $[S]_D$ equals the coefficient of $(\theta^T \epsilon \theta)^2$ in S plus a perfect space-time divergence. Hence, in terms of the Berezin integral, we can write

$$\int [f(\Phi)]_F d^4x = \int [D_R^2 \tilde{S}]_F d^4x = c_2 \int [\tilde{S}]_D d^4x = c_2 \int \tilde{S} d^4x d^4\theta$$

Now, we can write the total matter action as

$$\int (D_R^2 S)^*(D_R^2 S) d^4x d^4\theta + c_1 \int [f(D_R^2 S)]_F d^4x$$

We recall that from the definition of Majorana Fermionic parameters,

$$\theta^* = \gamma^5 \epsilon \gamma^0 \theta$$

where $\gamma^5 \epsilon \gamma^0$ has the block structure

$$\begin{pmatrix} 0 & e \\ -e & 0 \end{pmatrix}$$

and, we can interpret this equation in terms of 2×1 components as

$$\theta_L^* = e\theta_R, \theta_R^* = -e\theta_L$$

where

$$e^2 = -I_2$$

This is in agreement with the condition

$$(\theta_L^*)^* = \theta_L, (\theta_R^*)^* = \theta_R$$

since

$$e^* = e, e^2 = -I_2$$

provided that we interpret θ_L and θ_R as 2×1 Fermionic vectors. Note that

$$\epsilon = diag[e, e], \gamma^5\epsilon = diag[e, -e]$$

we can write

$$\bar{\theta} = (\theta^*)^T\gamma^0 = \theta^T\gamma^5\epsilon$$

and hence,

$$\bar{\theta}_L = \theta_L^T\gamma^5\epsilon, \bar{\theta}_R = \theta_R^T\gamma^5\epsilon$$

when interpreted as 4×1 Fermionic vectors. Also,

$$D^* = \gamma^5\epsilon\gamma^0 D$$

so that

$$\bar{D} = (D^*)^T\gamma^0 = -D^T\gamma^0\gamma^5\epsilon\gamma^0 = D^T\gamma^5\epsilon$$

and hence,

$$D_R^* = ((1-\gamma^5)/2)D^* = \gamma^5\epsilon\gamma^0 D_L,$$
$$D_L^* = ((1+\gamma^5)/2)D^* = \gamma^5\epsilon\gamma^0 D_R$$

when interpreted as 4×1 vector valued super-vector fields. When interpreted as 2×1 vector valued super vector fields, these equations should be read as

$$D_R^* = -eD_L, D_L^* = eD_R$$

when interpreted as 2×1 vector fields, it should be understood that D_L is represented by $\begin{pmatrix} D_L \\ 0 \end{pmatrix}$ and D_R by $\begin{pmatrix} 0 \\ D_R \end{pmatrix}$. Then,

$$\int [\Phi^*\Phi]_D d^4x = \int \Phi^*\Phi d^4x d^4\theta$$

$$= \int (D_R^2 S)^*(D_R^2 S) d^4x d^4\theta = \int \Phi^* D_R^2 S d^4x d^4\theta$$

$$= -\int (D_L^2\Phi^*) S d^4x d^4\theta$$

This is because

$$\Phi^* D_R^2 S = \Phi^* D_R^T\epsilon D_R S =$$
$$(-1)^{p(\Phi)} D_R^T(\Phi^*\epsilon D_R S) - (1)^{p(\Phi)} D_R^T\Phi^*.\epsilon D_R S$$
$$= (-1)^{p(\Phi)} D_R^T(\Phi^*\epsilon D_R S) + (1)^{p(\Phi)}(\epsilon D_R)^T\Phi^*.D_R S$$
$$= (-1)^{p(\Phi)} D_R^T(\Phi^*\epsilon D_R S) + D_R^T((\epsilon D_R)\Phi^*.S)$$
$$-(D_R^T\epsilon D_R\Phi^*).S$$

and

$$-(D_R^T\epsilon D_R\Phi^*)S == -(D_R^{*T}\epsilon D_R^*\Phi)^* S$$
$$= -(D_L^T(\gamma^5\epsilon\gamma^0)^T\epsilon\gamma^5\epsilon\gamma^0 D_L\Phi)^* S$$

$$= (D_L^T \gamma^0 \epsilon \gamma^5 \epsilon \gamma^5 \epsilon \gamma^0 D_L \Phi)^* S$$

$$= -(D_L^T \gamma^0 \epsilon \gamma^0 D_L \Phi)^* S =$$

$$= -(D_L^T \epsilon D_L \Phi)^* S = -(D_L^2 \Phi)^* S$$

Thus,

$$\int \Phi^* D_R^2 S d^4x d^4\theta = -\int (D_L^2 \Phi)^* S d^4x d^4\theta$$

and in exactly the same way, we can show that

$$\int [\Phi^* \delta\Phi]_D d^4x = \int [\Phi^* D_R^2 \delta S]_D d^4x$$

$$= \int \Phi^* D_R^2 \delta S d^4x d^4\theta = -\int (D_L^2 \Phi)^* \delta S d^4x d^4\theta$$

Likewise consider

$$\delta \int [f(D_R^2 S)]_F d^4x = \int [\delta f(D_R^2 S)]_F d^4x$$

Now consider for example

$$\delta(D_R^2 S) = D_R^2 \delta S,$$

$$\delta((D_R^2 S)^2) = D_R^2 \delta S . D_R^2 S + (-1)^{p(S)} D_R^2 S . D_R^2 \delta S$$

$$= D_R^2 \delta S . D_R^2 S + (-1)^{p(S)+p(S)p(\delta S)} + D_R^2 \delta S . D_R^2 S$$

$$= 2 D_R^2 \delta S . D_R^2 S$$

since

$$p(\delta S) = p(S)$$

Likewise, in general, we have

$$\delta (D_R^2 S)^n = D_R^2 \delta S . n (D_R^2 S)^{n-1}$$

and hence,

$$\delta f(D_R^2 S) = D_R^2 \delta S . f'(D_R^2 S) = D_R^2 \delta S . f'(\Phi)$$

Now,

$$D_R f'(\Phi) = D_R f'(D_R^2 S) = 0$$

since D_R operating on a constant is zero and the product of three $D_R's$ is zero. Thus,

$$D_R(\delta S . f'(\Phi)) = D_R \delta S . f'(\Phi),$$

$$D_R^2(\delta S . f'(\Phi)) = D_R^2 \delta S . f'(\Phi)$$

Thus,

$$\delta \int [f(\Phi)]_F d^4x = \int [D_R^2 \delta S . f'(\Phi)]_F d^4x =$$

$$\int [D_R^2(\delta S.f'(\Phi))]_F d^4x = \int [\delta S.f'(\Phi)]_D d^4x$$

$$= \int f'(\Phi)\delta S d^4x d^4\theta$$

since δS, S and $D_R^2 S$ have the same parity. Thus, our superfield equations

$$\delta[\int [\Phi^*\Phi]_D d^4x + c_1 \int [f(\Phi)]_F d^4x] = 0$$

for Φ a left Chiral field result in the superfield equations

$$-(D_L^2\Phi)^* + c_1 f'(\Phi) = 0$$

or equivalently

$$D_L^2 D_R^2 S = \bar{c}_1 f'(D_R^2 S)^*$$

Note that both sides of this equation are right Chiral superfields. More generally, if we replace the term $[\Phi^*\Phi]_D$ in the Lagrangian by $[K(\Phi^*, \Phi)]_D$ with $\Phi = D_R^2 S$, then the variation of the corresponding action integral w.r.t S is given by

$$\int [\delta K((D_R^2 S)^*, D_R^2 S)/\delta\Phi] D_R^2 \delta S d^4x d^4\theta$$

$$= -\int [D_R^2 \delta K(D_L^2 S^*, D_R^2 S)/\delta\Phi] \delta S d^4x d^4\theta$$

Note that in deriving this equation, we have used the fact that D_R acting on any superfield consists of a sum of terms having lesser that four Fermionic parameter products and terms that are total Bosonic space-time divergences. These terms cancel out when integrated over $d^4xd^4\theta$. So our superfield equations in this case generalize to

$$-D_R^2 \delta K(D_L^2 S^*, D_R^2 S)/\delta\Phi + c_1 f'(D_R^2 S) = 0$$

Both sides here are left Chiral superfields.

Supergravity:

Let ω_μ^{mn} denote the spinor connection of the gravitational field. Then if Γ_m are the Dirac matrices in four dimensions and e_μ^m is the tetrad basis of space time being used, the covariant derivative of a spinor field is defined by

$$D_\mu\psi = (\partial_\mu + (1/4)\omega_\mu^{mn}\Gamma_{mn})\psi$$

where

$$\Gamma_{mn} = [\Gamma_m, \Gamma_n]$$

The curvature tensor in spinor notation is

$$R_{\mu\nu} = [\partial_\mu + (1/4)\omega_{\mu\nu}^{mn}\Gamma_{mn}, \partial_\nu + (1/4)\omega_\nu^{rs}\Gamma_{rs}]$$

$$= (1/4)(\omega_{\nu,\mu}^{mn} - \omega_{\nu,\mu}^{mn})\Gamma_{mn}$$

$$+(1/16)\omega_\mu^{mn}\omega_\nu^{rs}[\Gamma_{mn},\Gamma_{rs}]$$

Now using the anticommutator

$$\{\Gamma_m,\Gamma_n\}=2\eta_{mn}$$

we can easily show that

$$[\Gamma_{mn},\Gamma_{rs}]=4(\eta_{ms}\Gamma_{nr}+\eta_{nr}\Gamma_{ms}-\eta_{mr}\Gamma_{ns}-\eta_{ns}\Gamma_{mr})$$

Thus

$$R_{\mu\nu}=$$

$$=(1/4)(\omega_{\nu,\mu}^{mn}-\omega_{\nu,\mu}^{mn})\Gamma_{mn}$$

$$+(1/4)\omega_\mu^{mn}\omega_\nu^{rs}(\eta_{ms}\Gamma_{nr}+\eta_{nr}\Gamma_{ms}-\eta_{mr}\Gamma_{ns}-\eta_{ns}\Gamma_{mr})$$

This can be expressed as

$$R_{\mu\nu}=(1/4)R_{\mu\nu}^{mn}\Gamma_{mn}$$

where

$$R_{\mu\nu}^{mn}=\omega_{\nu,\mu}^{mn}-\omega_{\nu,\mu}^{mn}-\omega_\mu^{rn}\omega_\nu^{ms}\eta_{rs}+\omega_\mu^{ms}\omega_\nu^{rn}\eta_{sr}+\omega_\mu^{sn}\omega_\nu^{rm}\eta_{sr}-\omega_\mu^{mr}\omega_\nu^{ns}\eta_{rs}$$

$$=$$

$$\omega_{\nu,\mu}^{mn}-\omega_{\nu,\mu}^{mn}+\eta_{rs}(-\omega_\mu^{rn}\omega_\nu^{ms}+\omega_\mu^{ms}\omega_\nu^{rn}+\omega_\mu^{sn}\omega_\nu^{rm}-\omega_\mu^{mr}\omega_\nu^{ns})$$

$$-\omega_{\nu,\mu}^{mn}-\omega_{\nu,\mu}^{mn}+2\eta_{rs}(\omega_\mu^{mr}\omega_\nu^{sn}-\omega_\nu^{mr}\omega_\mu^{ns})$$

It is easily shown that when the spinor connection ω_μ^{mn} for the gravitational field is appropriately chosen so that the Dirac equation in curved space-time remains invariant under both diffeormophisms and local Lorentz transformations, then the Riemann curvature tensor as defined usually in terms of the Christoffel connection symbols, coincides with $R_{\mu\nu\rho\sigma}=R_{\mu\nu}^{mn}e_{m\rho}e_{n\sigma}$. In particular, $R=R_{\mu\nu}^{mn}e_m^\mu e_n^\nu$ is the scalar curvature of space-time. The spinor connection ω_μ^{mn} is chosen so that the covariant derivative of the tetrad e_μ^n having one spinor index and one vector index is zero:

$$0=D_\nu e_\mu^n=e_{\mu,\nu}^n-\Gamma_{\mu\nu}^\alpha e_\alpha^n+\omega_\nu^{nm}e_{m\mu}$$

This is an algebraic equation for ω_μ^{mn} and is easily solved. However when there are spinor fields like the gravitino in addition to the gravitational field specified by the tetrad e_n^μ (ie, the graviton), then the definition of the spinor connection has to be modified and it is expressed in terms of both the graviton and the gravitino fields. This equation is obtained by first considering the supergravity Lagrangian in four space-time dimensions

$$c_1eR+i\bar{\chi}_\mu\Gamma^{\mu\nu\rho}D_\nu\chi_\rho$$

where χ_μ is a Majorana spinor having an additional vector index μ. The graviton tetrad field e_n^μ is Bosonic while the gravitino χ_μ is Fermionic. These are all considered in the quantum theory to be operator valued fields. Note the following self consistent definitions:

$$\Gamma^{\mu} = e^{\mu}_{n}\Gamma^{n}, \Gamma_{\mu} = g_{\mu\nu}\Gamma^{\nu} = \Gamma^{n}e_{n\mu}$$

$$\Gamma_{n} = \eta_{nm}\Gamma^{m}, e^{n}_{\mu}e_{n\nu} = g_{\mu\nu}, e^{\mu}_{n}e_{m\mu} = \eta_{nm},$$

$$e_{n\mu} = \eta_{nm}e^{m}_{\mu} = g_{\mu\nu}e^{\nu}_{n}$$

$$\Gamma^{\mu\nu\rho} = \Gamma^{\mu}\Gamma^{\nu\rho} + \Gamma^{\nu}\Gamma^{\rho\mu} + \Gamma^{\rho}\Gamma^{\mu\nu}$$

where

$$\Gamma^{\mu\nu} = [\Gamma^{\mu}, \Gamma^{\nu}]$$

Thus $\Gamma^{\mu\nu\rho}$ is obtained by antisymmetrizing the product $\Gamma^{\mu}\Gamma^{\nu}\Gamma^{\rho}$ over all the three indices. we can also clearly write

$$\Gamma^{\mu\nu} = e^{\mu}_{m}e^{\nu}_{n}\Gamma^{mn}, \Gamma^{\mu\nu\rho} = e^{\mu}_{m}e^{\nu}_{n}e^{\rho}_{k}\Gamma^{mnk}$$

In general, we can define

$$\Gamma^{\mu_1}..\Gamma^{\mu_k} = \sum_{\sigma\in S_n} sgn(\sigma)\Gamma^{\mu_{\sigma 1}}...\Gamma^{\mu_{\sigma k}}$$

This is obtained by totally antisymmetrizing the product $\Gamma^{\mu_1}...\Gamma^{\mu_k}$ over all its k indices. The basic property of a Majorana Fermionic operator field $\psi(x)$ is that apart from all its components anticommuting with each other, it has four components and satisfies

$$(\psi(x)^*)^T = \psi^T\Gamma^5\epsilon\Gamma^0$$

where if

$$\psi(x) = \begin{pmatrix} \psi_1(x) \\ \psi_2(x) \\ \psi_3(x) \\ \psi_4(x) \end{pmatrix}$$

then

$$\psi(x)^* = \begin{pmatrix} \psi_1(x)^* \\ \psi_2(x)^* \\ \psi_3(x)^* \\ \psi_4(x)^* \end{pmatrix}$$

$\psi_k(x)^*$ denoting the operator adjoint of $\psi_k(x)$ in the Fock space on which it acts. Also we define

$$\psi(x)^T = [\psi_1(x), \psi_2(x), \psi_3(x), \psi_4(x)]$$

so that we have

$$(\psi(x)^*)^T = [\psi_1(x)^*, \psi_2(x)^*, \psi_3(x)^*, \psi_4(x)^*]$$

Now observe that

$$\epsilon = diag[i\sigma^2, i\sigma^2], \sigma^2 = \begin{pmatrix} 0 & -i \\ i & 0 \end{pmatrix}$$

Note that ϵ is a real skewsymmetric matrix. we write

$$e = i\sigma^2 = \begin{pmatrix} 0 & 1 \\ -1 & 0 \end{pmatrix}$$

so that

$$\epsilon = diag[e, e], e^2 = -I$$

$$\Gamma^5 \epsilon \Gamma^0 = \begin{pmatrix} e & 0 \\ 0 & -e \end{pmatrix} \begin{pmatrix} 0 & I \\ I & 0 \end{pmatrix}$$

$$= \begin{pmatrix} 0 & e \\ -e & 0 \end{pmatrix}$$

Thus, the condition for ψ to be a Majorana Fermion can be stated as

$$\psi_{1:2}^* = e\psi_{3:4}, \psi_{3:4}^* = -e\psi_{1:2}$$

or equivalently,

$$(\psi_{1:2}^*)^T = -(\psi_{3:4}^*)^T e, (\psi_{3:4}^*)^T = (\psi_{1:2}^*)^T e$$

Also,

$$\Gamma^0\Gamma^n, \Gamma^0\Gamma^\mu, \Gamma^n\Gamma^0, \Gamma^\mu\Gamma^0$$

are Hermitian matrices. We observe that if ψ is a Majorana Fermion,

$$\bar{\psi} = (\psi^*)^T \Gamma^0 = \psi^T \Gamma^5 \epsilon \Gamma^0$$

So we can also write down the Lagrangian of the gravitino as

$$i\bar{\chi}_\mu \Gamma^{\mu\nu\rho} D_\nu \chi_\rho$$

$$= i\chi_\mu^{*T} \Gamma^0 \Gamma^{\mu\nu\rho} D_\nu \chi_\rho$$

$$= i\chi_\mu^T \Gamma^5 \epsilon \Gamma^0 \Gamma^{\mu\nu\rho} D_\nu \chi_\rho$$

We can verify that apart from a perfect divergence, this quantity is a Hermitian operator field. First observe that

$$(\Gamma^0\Gamma^\mu\Gamma^\nu\Gamma^\rho)^* =$$

$$(\Gamma^0\Gamma^\mu\Gamma^\nu\Gamma^0\Gamma^0\Gamma^\rho)^* =$$

$$\Gamma^0\Gamma^\rho\Gamma^\nu\Gamma^0\Gamma^0\Gamma^\mu$$

$$= \Gamma^0\Gamma^\rho\Gamma^\nu\Gamma^\mu$$

so that on antisymmetrizing over the three indices, we get

$$(\Gamma^0\Gamma^{\mu\nu\rho})^* = -\Gamma^0\Gamma^{\mu\nu\rho}$$

Thus,

$$(i\chi_\mu^{*^T}\Gamma^0\Gamma^{\mu\nu\rho}\partial_\nu\chi_\rho)^*$$
$$= i\chi_{\rho,\nu}^{*T}\Gamma^0\Gamma^{\mu\nu\rho}\chi_\mu$$
$$= i\chi_{\mu,\nu}^{*T}\Gamma^0\Gamma^{\rho\nu\mu}\chi_\rho$$
$$= -i\chi_{\mu,\nu}^{*T}\Gamma^0\Gamma^{\mu\nu\rho}\chi_\rho$$
$$= \partial_\nu(-i\chi_\mu^{*T}\Gamma^0\Gamma^{\mu\nu\rho}\chi_\rho)$$
$$+i\chi_\mu^{*T}\Gamma^0\Gamma^{\mu\nu\rho}\chi_{\rho,\nu}$$

proving our claim provided that we replace D_ν by ∂_ν. If we take the connection into account, ie

$$D_\nu\chi_\rho = \partial_\nu\chi_\rho + (1/4)\omega_\nu^{mn}\Gamma_{mn}\chi_\rho$$
$$-\Gamma^\alpha_{\rho\nu}\chi_\alpha$$

then it follows that we must prove the skew-Hermitianity of the operator fields

$$\chi_\mu^{*T}\Gamma^0\Gamma^{\mu\nu\rho}\Gamma_{mn}\chi_\rho.\omega_\nu^{mn} --- (a)$$

and

$$\chi_\mu^{*T}\Gamma^0\Gamma^{\mu\nu\rho}\chi_\alpha.\Gamma^\alpha_{\rho\nu} --- (b)$$

However the field (b) is identically zero since $\Gamma^{\mu\nu\rho}$ is antisymmetric in (ν, ρ) while $\Gamma^\alpha_{\rho\nu}$ is symmetric in (ν, ρ). Hence, we have to prove only the skew-Hermitianity of the field

$$\chi_\mu^{*T}\Gamma^0\Gamma^{\mu\nu\rho}\Gamma_{mn}\chi_\rho --- (c)$$

Now,

$$\Gamma^{\mu\nu\rho}\Gamma_{mn} = (1/2)[\Gamma^{\mu\nu\rho}, \Gamma_{mn}] + (1/2)\{\Gamma_{mn}\Gamma^{\mu\nu\rho}\}$$

and

$$[\Gamma_{pqr}, \Gamma_{mn}] = [\Gamma_p\Gamma_{qr} + \Gamma_q\Gamma_{rp} + \Gamma_r\Gamma_{pq}, \Gamma_{mn}]$$

Now,

$$[\Gamma_p\Gamma_{qr}, \Gamma_{mn}] = \Gamma_p[\Gamma_{qr}, \Gamma_{mn}] + [\Gamma_p, \Gamma_{mn}]\Gamma_{qr}$$
$$= 4\Gamma_p(\eta_{qn}\Gamma_{rm} + \eta_{rm}\Gamma_{qn} - \eta_{qm}\Gamma_{rn} - \eta_{rn}\Gamma_{qm})$$
$$+4(\eta_{pm}\Gamma_n - \eta_{pn}\Gamma_m)\Gamma_{qr}$$

Summing this equation over cyclic permutations of (pqr) gives us

$$[\Gamma_{pqr}, \Gamma_{mn}] =$$

$$4\sum_{(pqr)}\eta_{mq}(\Gamma_p\Gamma_{nr} + \Gamma_r\Gamma_{pn} + \Gamma_n\Gamma_{rp})$$

$$+4\sum_{(pqr)}\eta_{nq}(\Gamma_p\Gamma_{rm}+\Gamma_r\Gamma_{mp}+\Gamma_m\Gamma_{pr})$$

$$=4\sum_{(pqr)}(\eta_{mq}\Gamma_{pnr}+\eta_{nq}\Gamma_{prm})$$

Note that this quantity is antisymmetric w.r.t interchange of (m,n). It thus follows that

$$e_{q\nu}\chi_\mu^{*T}\Gamma^0[\Gamma^{\mu\nu\rho},\Gamma_{mn}]\chi_\rho$$

$$=\chi^{p*T}\Gamma^0[\Gamma_{pqr},\Gamma_{mn}]\chi^r$$

$$=4\sum_{(pqr)}[\eta_{mq}\chi^{p*T}\Gamma^0\Gamma_{pnr}\chi^r+\eta_{nq}\chi^{p*T}\Gamma^0\Gamma_{prm}\chi^r]$$

Now,

$$(\chi^{p*T}\Gamma^0\Gamma_{pnr}\chi^r)^*=$$

$$\chi^{r*T}\Gamma^0\Gamma^{rnp}\chi^p$$

$$=\chi^{p*T}\Gamma^0\Gamma^{pnr}\chi^r$$

which proves the Hermitianity of $\chi^{p*}\Gamma^0\Gamma_{pnr}\chi^r$ and hence of $\bar{\chi}_\mu[\Gamma^{\mu\nu\rho},\Gamma_{mn}]\chi_\rho$. In fact this quantity is identically zero. To see this, we use the Majorana Fermion property of χ^p to write

$$\chi^{p*T}\Gamma^0\Gamma_{pnr}\chi^r=$$

$$\chi^{pT}\Gamma^5\epsilon\Gamma_{pnr}\chi^r$$

and use the fact that

$$(\Gamma^5\epsilon\Gamma_{pnr})^T=-\Gamma_{pnr}^T\epsilon\Gamma^5$$

$$=-\epsilon\Gamma_{rnp}\Gamma^5=\Gamma^5\epsilon\Gamma_{rnp}=-\Gamma^5\epsilon\Gamma_{pnr}$$

where we have used the identities

$$\Gamma_n^T\epsilon=\epsilon\Gamma_n,\Gamma_n\Gamma^5=-\Gamma^5\Gamma_n$$

Then from the anticommutativity of the $\chi^{p's}$, we get

$$\chi^{pT}\Gamma^5\epsilon\Gamma_{pnr}\chi^r=$$

$$=-\chi^{rT}(\Gamma^5\epsilon\Gamma_{pnr})^T\chi^p=$$

$$=\chi^{rT}\Gamma^5\epsilon\Gamma_{pnr}\chi^p$$

$$=\chi^{pT}\Gamma^5\epsilon\Gamma_{rnp}\chi^r=-\chi^{pT}\Gamma^5\Gamma_{pnr}\chi^r$$

from which we conclude that

$$\chi^{pT}\Gamma^5\epsilon\Gamma_{rnp}\chi^r=0$$

Now consider

$$X=\chi_\mu^{*T}\Gamma^0\{\Gamma^{\mu\nu\rho},\Gamma_{mn}\}\chi_\rho$$

where $\{.,.\}$ denotes anticommutator. We have

$$X = X_1 + X_2$$

where

$$X_1 = \chi_\mu^{*T}\Gamma^0\Gamma^{\mu\nu\rho}\Gamma_{mn}\chi_\rho$$

$$X_2 = \chi_\mu^{*T}\Gamma^0\Gamma_{mn}\Gamma^{\mu\nu\rho}\chi_\rho$$

we have

$$X_1^* = \chi_\rho^{*T}\Gamma^0\Gamma_{mn}\Gamma^{\mu\nu\rho}\chi_\mu$$

$$= -\chi_\mu^{*T}\Gamma^0\Gamma_{mn}\Gamma^{\mu\nu\rho}\chi_\rho = -X_2$$

which shows that

$$X^* = -X$$

ie, X is skew Hermitian. Note that we have used the fact that

$$(\Gamma^0\Gamma_p\Gamma_q\Gamma_r\Gamma_m\Gamma_n)^* =$$

$$(\Gamma^0\Gamma_p\Gamma^0\Gamma^0\Gamma_q\Gamma_r\Gamma^0\Gamma^0\Gamma_m\Gamma^0\Gamma^0\Gamma_n)^* =$$

$$\Gamma^0\Gamma_n\Gamma_m\Gamma^0\Gamma^0\Gamma_r\Gamma^0\Gamma^0\Gamma_q\Gamma_p\Gamma^0\Gamma^0 =$$

$$\Gamma^0\Gamma_n\Gamma_m\Gamma_r\Gamma_q\Gamma_p$$

since $\Gamma^0\Gamma_n, \Gamma_n\Gamma^0, \Gamma^0$ are Hermitian and $\Gamma^{02} = I$. Thus, by antisymmetrizing over (pqr) and over (mn), we get

$$(\Gamma^0\Gamma_{pqr}\Gamma_{mn})^* = \Gamma^0\Gamma_{nm}\Gamma_{rqp}$$

$$= \Gamma^0\Gamma_{mn}\Gamma_{pqr}$$

This proves that

$$i\bar{\chi}_\mu\Gamma^{\mu\nu\rho}D_\nu\chi_\rho$$

is a Hermitian operator field.

Now consider the following local supersymmetry transformation

$$\delta\chi_\mu(x) = D_\mu\epsilon(x), \delta e_\mu^n = K\bar{\epsilon}(x)\Gamma^n\chi_\mu(x)$$

where $\epsilon(x)$ is an infinitesimal Majorana Fermionic parameter. We can easily check that $D_\mu\epsilon$ also satisfies the Majorana Fermion property. Indeed,

$$((D_\mu\epsilon(x))^*)^T = \partial_\mu(\epsilon(x)^*)^T + (\epsilon(x)^*)^T\Gamma_{mn}^*\omega_\mu^{mn}$$

$$= (\partial_\mu\epsilon(x))^T\Gamma^5\epsilon\Gamma^0 + \epsilon(x)^T\Gamma^5\epsilon\Gamma^0\Gamma_{mn}^*\omega_\mu^{mn}$$

$$= (\partial_\mu\epsilon(x))^T\Gamma^5\epsilon\Gamma^0 - \epsilon(x)^T\Gamma^5\epsilon\Gamma_{mn}\Gamma^0\omega_\mu^{mn}$$

Now,

$$\epsilon\Gamma_{mn} = -\Gamma_{mn}^T\epsilon$$

since

$$\Gamma_n^T \epsilon = \epsilon \Gamma_n$$

Thus, since $\Gamma^{0T} = \Gamma^0$, it follows that

$$\Gamma^5 \epsilon \Gamma_{mn} \Gamma^0 = \Gamma^5 \epsilon \Gamma_{mn} \Gamma^0 =$$

$$-\Gamma^5 \Gamma_{mn}^T \epsilon \Gamma^0 = -\Gamma_{mn}^T \Gamma^5 \epsilon \Gamma^0$$

This gives

$$((D_\mu \epsilon(x))^*)^T =$$

$$(\partial_\mu \epsilon(x))^T \Gamma^5 \epsilon \Gamma^0$$

$$+\epsilon(x)^T \Gamma_{mn}^T \Gamma^5 \epsilon \Gamma^0$$

$$= (D_\mu \epsilon(x))^T \Gamma^5 \epsilon \Gamma^0$$

proving thereby the Majorana property of $D_\mu \epsilon(x)$. Now under the local supersymmetry transformation of χ_μ, the Gravitino Lagrangian changes by

$$\delta_\chi(\bar{\chi}_\mu \Gamma^{\mu\nu\rho} D_\nu \chi_\rho) =$$

$$= \delta \bar{\chi}_\mu \Gamma^{\mu\nu\rho} D_\nu \chi_\rho +$$

$$\bar{\chi}_\mu \Gamma^{\mu\nu\rho} D_\nu \delta \chi_\rho$$

$$= \bar{D}_\mu \epsilon(x) \Gamma^{\mu\nu\rho} D_\nu \chi_\rho$$

$$+\bar{\chi}_\mu \Gamma^{\mu\nu\rho} D_\nu D_\rho \epsilon(x)$$

The term in this quantity that is quadratic in $\{\omega_\mu^{mn}\}$ is given by

$$-\omega_\mu^{mn} \omega_\rho^{rs} [\epsilon(x)^T \Gamma_{mn}^T \Gamma^5 \epsilon \Gamma^{\mu\nu\rho} \Gamma_{rs} \chi_\nu$$

$$+\chi_\nu^T \Gamma^5 \epsilon \Gamma^{\mu\nu\rho} \Gamma_{mn} \Gamma_{rs} \epsilon(x)]$$

we must first prove that this is Hermitian under the assumption that $\epsilon(x)$ and $\chi_\mu(x)$ are mutually anticommuting Majorana Fermionic fields. Note that we can also express this as

$$-\omega_\mu^{mn} \omega_\rho^{rs} [\epsilon(x)^{*T} \Gamma_{mn}^* \Gamma^0 \Gamma^{\mu\nu\rho} \Gamma_{rs} \chi_\nu(x)$$

$$+\chi_\nu(x)^{*T} \Gamma^0 \Gamma^{\mu\nu\rho} \Gamma_{mn} \Gamma_{rs} \epsilon(x)]$$

Now,

$$\Gamma_{mn}^* \Gamma^0 = -\Gamma^0 \Gamma_{mn}$$

Thus,

$$\Gamma_{mn}^* \Gamma^0 \Gamma_{pqk} \Gamma_{rs} =$$

$$-\Gamma^0 \Gamma_{mn} \Gamma_{pqk} \Gamma_{rs}$$

Thus,

$$(\epsilon(x)^{*T} \Gamma_{mn}^* \Gamma^0 \Gamma^{\mu\nu\rho} \Gamma_{rs} \chi_\nu)^* =$$

$$-(\epsilon(x)^{*T}\Gamma^0\Gamma_{mn}\Gamma^{\mu\nu\rho}\Gamma_{rs}\chi_\nu)^* =$$

$$\chi_\nu^*\Gamma^0\Gamma_{rs}\Gamma^{\mu\nu\rho}\Gamma_{mn}\epsilon(x)$$

$$= \chi_\nu^T\Gamma^5\epsilon\Gamma_{rs}\Gamma^{\mu\nu\rho}\Gamma_{mn}\epsilon(x)$$

$$= -\epsilon(x)^T(\Gamma^5\epsilon\Gamma_{rs}\Gamma^{\mu\nu\rho}\Gamma_{mn})^T\chi_\nu$$

$$= \epsilon(x)^T\Gamma_{mn}^T(\Gamma^{\mu\nu\rho})^T\Gamma_{rs}^T\Gamma^5\epsilon\chi_\nu$$

$$= \epsilon(x)^T\Gamma^5\epsilon\Gamma_{mn}\Gamma^{\mu\nu\rho}\Gamma_{rs}\chi_\nu$$

$$= (\epsilon(x)^{*T}\Gamma^0\Gamma_{mn}\Gamma^{\mu\nu\rho}\Gamma_{rs}\chi_\nu$$

$$= -\epsilon(x)^{*T}\Gamma_{mn}^*\Gamma^0\Gamma^{\mu\nu\rho}\Gamma_{rs}\chi_\nu$$

This proves that $(\epsilon(x)^{*T}\Gamma_{mn}^*\Gamma^0\Gamma^{\mu\nu\rho}\Gamma_{rs}\chi_\nu)$ is skew-Hermitian. Note that if we replace $\epsilon(x)$ by $i\epsilon(x)$ where $\epsilon(x)$ is a Majorana Fermion, then the above quantity becomes Hermitian. Consider now the second term. It is

$$\chi_\nu^T\Gamma^5\epsilon\Gamma^{\mu\nu\rho}\Gamma_{mn}\Gamma_{rs}\epsilon(x)]$$

$$= \chi_\nu^*\Gamma^0\Gamma^{\mu\nu\rho}\Gamma_{mn}\Gamma_{rs}\epsilon(x)$$

$$= \chi_\nu^*\Gamma^0[\Gamma^{\mu\nu\rho},\Gamma_{mn}]\Gamma_{rs}\epsilon(x)$$

$$+\chi_\nu^*\Gamma^0\Gamma_{mn}\Gamma^{\mu\nu\rho}\Gamma_{rs}\epsilon(x)$$

This is also easily shown to be skew-Hermitian. Indeed, its adjoint is given by

The general theory of Chiral superfields: The general superfield has the form

$$S(x,\theta) = C(x) + \theta^T\epsilon\omega(x) + \theta^T\epsilon\theta M(x) + \theta^T\gamma^5\epsilon N(x) + \theta^T\epsilon\gamma^\mu\theta V_\mu(x)$$

$$\theta^T\epsilon\theta.\theta^T\gamma^5\epsilon(\lambda(x) + a.\gamma^\nu\omega_{,\nu}(x) + (\theta^T\epsilon\theta)^2(D(x) + b.\Box C(x))$$

Under the above mentioned supersymmetry transformation

$$\alpha^T\gamma^5\epsilon.L = \bar{\alpha}L$$

where

$$L = \gamma^\mu\theta\partial_\mu + \gamma^5\epsilon\partial_\theta$$

the change in the superfield S is given by

$$\delta S = \bar{\alpha}LS$$

Here α is a Majorana Fermionic parameter. One can compute the change in the component fields $C, \omega, M, N, V_\mu, \lambda, D$ under this infinitesimal supersymmetry transformation and show that D changes by a perfect space-time four divergence and hence can be used as a candidate Lagrangian. However, in the case when the superfield is such that $\lambda = 0, D = 0, V_\mu = B_{,\mu}$, the resulting superfield is called Chiral and it is easy to prove that under a supersymmetry transformation, a Chiral superfield transforms into a Chiral superfield. The general Chiral superfield can be expressed as

$$S_c(x,\theta) = C(x) + \theta^T \epsilon\omega(x) + \theta^T \epsilon\theta M(x) + \theta^T \gamma^5 \epsilon\theta N(x) + \theta^T \epsilon\gamma^\mu \theta B_{,\mu}(x) +$$

$$a\theta^T \epsilon\theta.\theta^T \gamma^5 \epsilon\gamma^\nu \omega_{,\nu}(x) + b(\theta^T \epsilon\theta)^2 \Box C(x) ---(1)$$

Let us first prove that the class of Chiral supefields is supersymmetry invariant. To do so, we first note that for the superfield S_c above, the change in λ under the above infinitesimal supersymmetry transformation, obtained by equating the cubic terms in θ, is given by

$$\theta^T \epsilon\theta.\theta^T \gamma^5 \epsilon(\delta\lambda + \gamma^\nu \delta\omega_{,\nu})$$

$$= \bar{\alpha}\gamma^\mu \theta(\theta^T \epsilon\theta.M_{,\mu} + \theta^T \gamma^5 \epsilon\theta N_{,\mu} + \theta^T \epsilon\gamma^\nu \theta.B_{,\nu\mu})$$

$$+ \bar{\alpha}\gamma^5 \epsilon(4b.\theta^T \epsilon\theta.\epsilon\theta)\Box C$$

The change in ω is given by equating linear terms in θ:

$$\theta^T \epsilon.\delta\omega = \bar{\alpha}\gamma^\mu \theta C_{,\mu} +$$

$$\bar{\alpha}\gamma^5 \epsilon(2\epsilon\theta.M + 2\gamma^5 \epsilon N + 2\epsilon\gamma^\nu \theta.B_{,\nu})$$

By using the identities

$$\epsilon\gamma^\mu, \epsilon, \gamma^5 \epsilon$$

are skewsymmetric,

$$\theta\theta^T = (1/4)(\theta^T \epsilon\theta.\epsilon + \theta^T \gamma^5 \epsilon\gamma^5 \epsilon + \theta^T \epsilon\gamma^\mu \theta.\epsilon\gamma_\mu) --- (a)$$

and

$$\{\gamma^\mu, \gamma^\nu\} = 2\eta^{\mu\nu}$$

it is easy to see from the above relations that

$$\delta\lambda = 0$$

Likewise we can verify that $\delta D = 0$ and the condition that $V_\mu = B_{,\mu}$, ie, V_μ is a perfect four gradient also remains invariant under a supersymmetry transformation. Indeed,

$$(\theta^T \epsilon\theta)^2(\delta D + b.\Box\delta C)$$

$$= \bar{\alpha}\gamma^\mu \theta(\theta^T \epsilon\theta).\theta^T \gamma^5 \epsilon(a.\gamma^\nu \omega_{,\mu\nu})$$

and

$$\delta C = \bar{\alpha}\gamma^5 \epsilon\partial_\theta \theta^T \epsilon\omega$$

$$= \alpha^T \epsilon\omega$$

together imply that

$$\delta D = 0$$

Note that the product of any two distinct members of the six quantities

$$\theta^T\epsilon\theta, \theta^T\gamma^5\epsilon\theta, \theta^T\epsilon\gamma^\mu\theta$$

is zero and therefore

$$\theta(\theta^T\epsilon\theta)\theta^T = (\theta^T\epsilon\theta)\epsilon/4$$

We also use

$$\gamma^\mu\gamma^\nu\omega_{,\mu\nu} = \eta^{\mu\nu}\omega_{,\mu\nu} = \Box\omega$$

Finally, we must verify that $V_\mu = B_{,\mu}$ changes by an exact four gradient under a supersymmetry transformation. To see this we use the equations corresponding to quadratic terms in the θ:

$$\theta^T\epsilon\theta.\delta M + \theta^T\gamma^5\epsilon\theta.\delta N + \theta^T\epsilon\gamma^\mu\theta\delta V_\mu =$$

$$\bar{\alpha}\gamma^\mu\theta\theta^T\epsilon\omega_{,\mu} + \alpha^T\partial_\theta(\theta^T\epsilon\theta.\theta^T\gamma^5\epsilon(a.\gamma^\mu\omega_{,\mu})$$

$$= \alpha^T\gamma^5\epsilon\gamma^\mu\theta\theta^T\epsilon\omega_{,\mu} +$$

$$\alpha^T(2\epsilon\theta\theta^T + \theta^T\epsilon\theta\gamma^5\epsilon)(a\gamma^\mu\omega_{,\mu})$$

Noting that the terms $\theta^T\epsilon\theta, \theta^T\gamma^5\epsilon\theta, \theta^T\epsilon\gamma^\mu\theta$ are all the six linearly independent quadratic combinations of the θ, we get on equating the coefficients of $\theta^T\epsilon\gamma^\mu\theta$ on both sides after recalling identity (a) that

$$\delta V_\mu = \alpha^T\gamma^5\epsilon\gamma^\nu(1/4)\epsilon\gamma_\mu\omega_{,\nu}$$

$$+(a/2)\alpha^T\epsilon.\epsilon\gamma_\mu\gamma^\nu\omega_{,\nu}$$

which indeed proves that δV_μ is a perfect four gradient thereby completing the proof that the class of Chiral fields is closed under supersymmetry transformations.

Remark: Matter fields in non-Abelian quantum field theory get generalized in supersymmetry theory to super matter fields which are obtained from the D-component of products of left Chiral fields with their complex conjugates while gauge fields in quantum field theory get generalized to super-gauge fields which are derived from the gauge fields V_μ, the gaugino fields λ and the auxiliary fields D. In conventional non-Abelian quantum field theory, the matter fields transform according to a representation of the gauge group with the group element being in general local, ie, a function of the space-time coordinates while the gauge fields transform according to the adjoint representation of the gauge group plus an additional factor involving space-time gradients of the representation of the local gauge group elements. In supersymmetry theory, the matter field Lagrangian derived from D-component of quadratic combinations of left Chiral fields and their complex conjugates comprises of the scalar field part, the Dirac spinor field part and and auxiliary part which is determined in terms of the first two parts by setting the variational derivative of the corresponding action w.r.t it to zero, ie, it is determined by its field equation which is a purely algebraic equation for it. Supersymmetry predicts then that the mass term in the Dirac field component Lagrangian contains a mass term that depends on the scalar field.

Thus an arbitrary Chiral superfield has the form (1) which we repeat here for convenience

$$S_c(x,\theta)=C(x)+\theta^T\epsilon\omega(x)+\theta^T\epsilon\theta M(x)+\theta^T\gamma^5\epsilon\theta N(x)+\theta^T\epsilon\gamma^\mu\theta B_{,\mu}(x)+$$

$$a\theta^T\epsilon\theta.\theta^T\gamma^5\epsilon\gamma^\nu\omega_{,\nu}(x)+b(\theta^T\epsilon\theta)^2\Box C(x)---(1)$$

Now,

$$\theta^T\epsilon\theta=\theta_R^T\epsilon\theta_R+\theta_L^T\epsilon\theta_L$$

$$\theta^T\gamma^5\epsilon\theta=\theta_R^T\gamma^5\epsilon\theta_R+\theta_L^T\gamma^5\epsilon\theta_L$$

$$=-\theta_R^T\epsilon\theta_R+\theta_L^T\epsilon\theta_L$$

since

$$\theta_R^T\epsilon\theta_L=0,\theta_R^T\gamma^5\epsilon\theta_L=0$$

and

$$\theta=\theta_R+\theta_L=(1-\gamma^5)\theta/2+(1+\gamma^5)\theta/2$$

Also since

$$\theta_R^T\epsilon\gamma^\mu\theta_R=\theta_L^T\epsilon\gamma^\mu\theta_L=0$$

it follows that

$$\theta^T\epsilon\gamma^\mu\theta=2\theta_R^T\epsilon\gamma^\mu\theta_L$$

$$=2\theta_L^T\epsilon\gamma^\mu\theta_R$$

since θ_R and θ_L anticommute and $\epsilon\gamma^\mu$ is skew-symmetric. Further,

$$\theta^T\epsilon\theta.\theta^T\gamma^5\epsilon=$$

$$(\theta_R^T\epsilon\theta_R+\theta_L^T\epsilon\theta_L)(\theta_R^T+\theta_L^T)\gamma^5\epsilon$$

$$=\theta_R^T\epsilon\theta_R.\theta_L^T\gamma^5\epsilon+\theta_L^T\epsilon\theta_L\theta_R^T\gamma^5\epsilon$$

Now observe that

$$\theta_R\theta_L^T=(1/4)(1-\gamma^5)\theta\theta^T(1+\gamma^5)$$

$$=(1/16)(1-\gamma^5)[\theta^T\epsilon\theta\epsilon+\theta^T\gamma^5\epsilon\theta\gamma^5\epsilon+\theta^T\epsilon\gamma^\mu\theta\epsilon\gamma_\mu](1+\gamma^5)$$

$$=(1/8)(1-\gamma^5)\epsilon\gamma_\mu(1+\gamma^5)\theta_R^T\epsilon\gamma^\mu\theta_L$$

$$=(1/4)(1-\gamma^5)\epsilon\gamma_\mu\theta_R^T\epsilon\gamma^\mu\theta_L$$

and likewise,

$$\theta_L.\theta_R^T=(1/4)(1+\gamma^5)\epsilon\gamma_\mu\theta_R^T\epsilon\gamma^\mu\theta_L$$

Thus, using the fact that

$$\theta_R^T(1-\gamma^5)/2=\theta_R^T,\theta_L^T(1+\gamma^5)/2=\theta_L$$

we get

$$\theta^T\epsilon\theta.\theta^T=$$

$$\theta_R^T\epsilon\theta_R^T\epsilon\gamma^\mu\theta_L\epsilon\gamma_\mu/2$$

$$+\theta_L^T\epsilon\theta_L^T\epsilon\gamma^\mu\theta_R\epsilon\gamma_\mu/2$$

$$= -\theta_R^T \epsilon\gamma^\mu \theta_L \theta_R^T \gamma_\mu/2$$
$$-\theta_R^T \epsilon\gamma^\mu \theta_L \theta_L^T \gamma_\mu/2$$

Another way to see this is as follows:

$$\theta^T \epsilon\theta\theta^T =$$

$$(\theta_R^T + \theta_L^T)\epsilon(\theta^T \epsilon\theta\epsilon + \theta^T \gamma^5 \epsilon\theta\gamma^5 \epsilon + \theta^T \epsilon\gamma^\mu \theta.\epsilon\gamma_\mu)/4$$
$$= (\theta_R^T + \theta_L^T)\epsilon(\theta_R^T \epsilon\theta_R \epsilon + \theta_L^T \epsilon\theta_L \epsilon + \theta_R^T \gamma^5 \epsilon\theta_R \gamma^5 \epsilon + \theta_L^T \gamma^5 \epsilon\theta_L \gamma^5 \epsilon + 2\theta_R^T \epsilon\gamma^\mu \theta_L.\epsilon\gamma_\mu)/4$$

Now,

$$\theta_R^T \gamma^5 \epsilon\theta_R = -\theta_R^T \epsilon\theta_R,$$
$$\theta_L^T \gamma^5 \epsilon\theta_L = \theta_L^T \epsilon\theta_L$$

and hence

$$\theta_R^T \epsilon(\theta_L^T \epsilon\theta_L \epsilon + \theta_L^T \gamma^5 \epsilon\theta_L \gamma^5 \epsilon)$$
$$= \theta_R^T \epsilon(\theta_L^T \epsilon\theta_L (1 + \gamma^5)\epsilon/2$$
$$= -\theta_R^T (1 + \gamma^5)\theta_L^T \epsilon\theta_L/2 = 0$$

and likewise,

$$\theta_L^T \epsilon(\theta_R^T \epsilon\theta_R \epsilon + \theta_R^T \gamma^5 \epsilon\theta_R \gamma^5 \epsilon)$$
$$= -\theta_L^T \theta_R^T \epsilon\theta_R (1 - \gamma^5)/2 = 0$$

Thus, we get

$$\theta^T \epsilon\theta\theta^T = -(1/2)\theta_R^T \gamma_\mu \theta_R^T \epsilon\gamma^\mu \theta_L - (1/2)\theta_L^T \gamma_\mu \theta_R^T \epsilon\gamma^\mu \theta_L$$

Thus, the Chiral superfield (1) can be expressed as

$$S_c(x, \theta) =$$

$$C(x) + (\theta_L^T \epsilon\omega(x) + \theta_R^T \epsilon\omega(x)) +$$
$$(\theta_L^T \epsilon\theta_L + \theta_R^T \epsilon\theta_R)M(x)$$
$$+(\theta_L^T \epsilon\theta_L - \theta_R^T \epsilon\theta_R)N(x)$$
$$+2\theta_R^T \epsilon\gamma^\mu \theta_L B_{,\mu}(x)$$
$$-(a/2)(\theta_R^T \epsilon\gamma^\mu \theta_L)(\theta_L^T \gamma_\mu \gamma^5 \epsilon\gamma^\nu \omega_{,\nu} + \theta_R^T \gamma_\mu \gamma^5 \epsilon\gamma^\nu \omega_{,\nu})$$
$$+(1/2)(\theta_R^T \epsilon\gamma^\mu \theta_L).(\theta_R^T \epsilon\gamma_\mu \theta_L)b\Box C$$

An alternate more convenient formula for the cubic term is obtained as follows: We've already noted that

$$\theta^T \epsilon\theta.\theta^T =$$

$$(\theta_R^T \epsilon\theta_R + \theta_L^T \epsilon\theta_L)(\theta_R^T + \theta_L^T)$$
$$= \theta_R^T \epsilon\theta_R.\theta_L^T + \theta_L^T \epsilon\theta_L.\theta_R^T$$

On the other hand, consider the expression

$$A = \theta_R^T \epsilon \gamma^\mu \theta_L . \theta_L^T \epsilon . \omega_{,\mu}$$

We have

$$\theta_L \theta_L^T = ((1+\gamma^5)/2)\theta\theta^T(1+\gamma^5)/2$$

$$= ((1+\gamma^5)/2)(\theta^T \epsilon\theta\epsilon + \theta^T\gamma^5\epsilon\theta\gamma^5\epsilon)((1+\gamma^5)/2)(1/4)$$

$$= \theta^T(1+\gamma^5)\epsilon\theta((1+\gamma^5)/8)$$

$$= \theta_L^T\epsilon\theta_L(1+\gamma^5)\epsilon/4$$

Thus,

$$A = (1/4)\theta_L^T\epsilon\theta_L.\theta_R^T\epsilon\gamma^\mu(1+\gamma^5)\epsilon\epsilon\omega_{,\nu}$$

$$= -(1/2)\theta_L^T\epsilon\theta_L.\theta_R^T\epsilon\gamma^\mu\omega_{,\mu}$$

$$= (1/2)\theta_L^T\epsilon\theta_L.\theta_R^T\gamma^5\epsilon\gamma^\mu\omega_{,\mu}$$

Likewise, defining

$$B = \theta_R^T\epsilon\gamma^\mu\theta_L.\theta_R^T\epsilon.\omega_{,\mu}$$

$$= \theta_L^T\epsilon\gamma^\mu\theta_R.\theta_R^T\epsilon\omega_{,\mu}$$

we get using

$$\theta_R\theta_R^T = ((1-\gamma^5)/2)\theta\theta^T((1+\gamma^5)/2)$$

$$= ((1-\gamma^5)/2)(\theta^T\epsilon\theta\epsilon + \theta^T\gamma^5\epsilon\theta\gamma^5\epsilon)((1-\gamma^5)/2)(1/4)$$

$$= ((1-\gamma^5)/8)\epsilon.\theta^T(1-\gamma^5)\epsilon\theta$$

$$= ((1-\gamma^5)/4)\epsilon.\theta_R^T\epsilon\theta_R$$

and hence,

$$B = -\theta_L^T\epsilon\gamma^\mu((1-\gamma^5)/4)\omega_{,\mu}\theta_R^T\epsilon\theta_R$$

$$= (-1/2)\theta_R^T\epsilon\theta_R.\theta_L^T\gamma^5\epsilon\gamma^\mu\omega_{,\mu}$$

Thus, the cubic term in the above Chiral field is given by

$$a\theta^T\epsilon\theta.\theta^T\gamma^5\epsilon\gamma^\mu\omega_{,\mu}$$

$$= a(\theta_L^T\epsilon\theta_L\theta_R^T\gamma^5\epsilon\gamma^\mu\omega_{,\mu} + \theta_R^T\epsilon\theta_R\theta_L^T\gamma^5\epsilon\gamma^\mu\omega_{,\mu})$$

$$= 2a(\theta_R^T\epsilon\gamma^\mu\theta_L\theta_L^T\epsilon\omega_{,\mu}$$

$$-\theta_R^T\epsilon\gamma^\mu\theta_L\theta_R^T\epsilon\omega_{,\mu})$$

Combining these two identities, we can express the general Chiral super-field as

$$S_c(x,\theta) =$$

$$C(x) + (\theta_L^T\epsilon\omega(x) + \theta_R^T\epsilon\omega(x)) +$$

$$(\theta_L^T\epsilon\theta_L + \theta_R^T\epsilon\theta_R)M(x)$$

$$+(\theta_L^T\epsilon\theta_L - \theta_R^T\epsilon\theta_R)N(x)$$

$$+2\theta_R^T\epsilon\gamma^\mu\theta_L B_{,\mu}(x)$$
$$+2a\theta_R^T\epsilon\gamma^\mu\theta_L(\theta_L^T\epsilon\omega_{,\mu} - \theta_R^T\epsilon\omega_{,\mu})$$
$$+(1/2)(\theta_R^T\epsilon\gamma^\mu\theta_L).(\theta_R^T\epsilon\gamma_\mu\theta_L)b\Box C$$

Note:

$$(\theta_R^T\epsilon\gamma^\mu\theta_L).(\theta_R^T\epsilon\gamma_\mu\theta_L)$$
$$= \theta_R^T\epsilon\gamma^\mu\theta_L\theta_L^T\epsilon\gamma_\mu\theta_R$$

and

$$\theta_L\theta_L^T = (1/4)(1+\gamma^5)\theta.\theta^T(1+\gamma^5)$$
$$= (1/4)(\theta^T\epsilon\theta\epsilon(1+\gamma^5)^2 + \theta^T\gamma^5\epsilon\theta\gamma^5\epsilon(1+\gamma^5)^2)$$
$$= (1/2)\theta^T\epsilon(1+\gamma^5)\theta.(1+\gamma^5)\epsilon$$
$$= \theta_L^T\epsilon\theta_L(1+\gamma^5)\epsilon$$

and hence,

$$(\theta_R^T\epsilon\gamma^\mu\theta_L).(\theta_R^T\epsilon\gamma_\mu\theta_L)$$
$$= (\theta_R^T\epsilon\gamma^\mu(1+\gamma^5)\epsilon.\epsilon\gamma_\mu\theta_R).(\theta_L^T\epsilon\theta_L)$$
$$= 4(\theta_R^T\epsilon\theta_R).(\theta_L^T\epsilon\theta_L)$$

since γ^5 anticommutes with γ^μ and $\gamma^\mu\gamma_\mu = -2$. On the other hand,

$$(\theta^T\epsilon\theta)^2 = (\theta_L^T\epsilon\theta_L + \theta_R^T\epsilon\theta_R)^2$$
$$= 2(\theta_R^T\epsilon\theta_R).(\theta_L^T\epsilon\theta_L)$$

Exercise: Show that the general Chiral superfield $S_c(x,\theta)$ is expressible as the sum of a left Chiral superfield and a right Chiral superfield, where a left Chiral field is a function of θ_L and $x_+^\mu = x^\mu + \theta_R^T\epsilon\gamma^\mu\theta_L$ and conversely a right Chiral super-field is a function of θ_R and $x_-^{mu}x^\mu - \theta_R^T\epsilon\gamma^\mu\theta_R$. Specifically, any left Chiral super-field can be expressed as

$$\Phi(x,\theta) = \phi(x_+) + \theta_L^T\epsilon\psi(x_+) + \theta_L^T\epsilon\theta_L F(x)$$

where ψ is a left Chiral field and any right Chiral field can be expressed as

$$\eta(x,\theta) = \phi(x_-) + \theta_R^T\epsilon\psi(x_-) + \theta_R^T\epsilon\theta_R F(x)$$

where ψ is right Chiral. Note that there is no loss of generality in assuming that ψ is left Chiral in the former case and right Chiral in the latter since if ψ is arbitrary, then

$$\theta_L^T\epsilon\psi = \theta_L^T\epsilon((1+\gamma^5)/2)\psi$$
$$\theta_R^T\epsilon\psi = \theta_R^T\epsilon((1-\gamma^5)/2)\psi$$

since γ^5 commutes with ϵ. Note that

$$\theta_L^* = \gamma^5\epsilon\gamma^0\theta_R,$$

$$\theta_R^* = \gamma^5\epsilon\gamma^0\theta_L$$

It is easy to verify that

$$\gamma^5\epsilon\gamma^0$$

is a real matrix whose square is the identity and this confirms the requirement that

$$(\theta_L^*)^* = \theta_L, (\theta_R^*)^* = \theta_R$$

Note that by the definition of the Majorana Fermion,

$$\theta^* = \gamma^5\epsilon\gamma^0\theta$$

which gives

$$(\theta_L + \theta_R)^* = \theta_L^* + \theta_R^* = \gamma^5\epsilon\gamma^0(\theta_L + \theta_R)$$

from which the desired relationships follow on equating the first two components and the last two components. Now,

$$(\theta_R^T\epsilon\gamma^\mu\theta_L)^*$$

$$= (\theta_R^*)^T conj(\epsilon\gamma^\mu)\theta_L^*$$

$$- (\gamma^5\epsilon\gamma^0\theta_L)^T conj(\epsilon\gamma^\mu)\gamma^5\epsilon\gamma^0\theta_R$$

$$= \theta_L^T\gamma^5\epsilon\gamma^0.conj(\epsilon\gamma^\mu)\gamma^5\epsilon\gamma^0\theta_R$$

$$= -\theta_L^T\gamma^5\epsilon\gamma^0(\epsilon\gamma^\mu)^*\gamma^5\epsilon\gamma^0\theta_R$$

since $\epsilon\gamma^\mu$ is skew-symmetric. Now since $\gamma^5\epsilon\gamma^0$ is Hermitian and γ^μ and γ^5 anticommute (note that γ^5 and ϵ commute), it follows that

$$\gamma^5\epsilon\gamma^0(\epsilon\gamma^\mu)^*\gamma^5\epsilon\gamma^0$$

$$= (\gamma^5\epsilon\gamma^0\epsilon\gamma^\mu\gamma^5\epsilon\gamma^0)^*$$

$$(\epsilon\gamma^0\epsilon\gamma^\mu\epsilon\gamma^0)^*$$

Now,

$$\epsilon\gamma^0 = \gamma^0\epsilon$$

and further,

$$(\epsilon\gamma^0)^* = -\gamma^0\epsilon$$

and $\gamma^0\gamma^\mu$ is Hermitian. Thus, the above equals

$$-(\gamma^0\gamma^\mu\epsilon\gamma^0)^*$$

$$= \gamma^0\epsilon\gamma^0\gamma^\mu = \epsilon\gamma^\mu$$

Thus, we have proved that

$$(\theta_R^T \epsilon\gamma^\mu \theta_L)^* = -\theta_L^T \epsilon\gamma^\mu \theta_R$$
$$= -\theta_R^T \epsilon\gamma^\mu \theta_L$$

It follows that

$$(x_+^\mu)^* = x_-^\mu$$

and since

$$\theta_L^* = \gamma^5 \epsilon\gamma^0 \theta_R$$

with $\gamma^5\epsilon\gamma^0$ being non-singular, the conjugate of any function of (θ_L, x_+^μ) is a function of (θ_R, x_-^μ) and conversely. Therefore, the conjugate of any left Chiral superfield is a right Chiral superfield and conversely.

Construction of supersymmetric Lagrangians (actions) from Chiral superfields. We shall observe that when we construct the supersymmetric Lagrangian as $[\Phi^*\Phi]_D$, then we get terms corresponding to the kinetic parts of the Klein-Gordon and Dirac Lagrangians while when we construct the supersymmetric Lagrangian as $[f(\Phi)]_F$, we get extra inertial parts for these Lagrqangians. In particular, we will observe the remarkable fact that when after constructing the total Lagrangian by adding these two components, we write down the field equations for the auxiliary fields D, F, we will be able to eliminate these terms and will obtain a broken supersymmetric Lagrangian. In particular, the solution to the auxiliary field equations determine the masses of the Dirac particle. In this way, supersymmetry is able to explain in a natural way how the scalar field couples to the Dirac field giving rise to massive Dirac particles after supersymmetry is broken.

Consider now the Lagrangian

$$L_1 = [\Phi^*\Phi]_D$$

where

$$\Phi = \phi(x_+) + \theta_L^T \epsilon\psi(x_+) + \theta_L^T \epsilon\theta_L F(x)$$

Then,

$$\Phi^* = \phi^*(x_-) + \theta_R^T \gamma^5 \epsilon\gamma^0 \epsilon\psi^*(x_-) + \theta_R^T \gamma^5 \epsilon\gamma^0 \epsilon\gamma^5 \epsilon\gamma^0 \theta_R F^*(x)$$
$$= \phi^*(x_-) + \theta_R^T \gamma^0 \psi^*(x) + \theta_R^T \epsilon\theta_R F^*(x)$$

We shall now evaluate all the terms in $[\Phi^*\Phi]_D$: First,

$$[\phi^*(x_-)\phi(x_+)]_D = T_1 + T_2 + T_3$$

where

$$T_1 = -\phi^*_{,\mu}\phi_{,\nu}(\theta_R^T \epsilon\gamma^\mu \theta_L).(\theta_R^T \epsilon\gamma^\nu \theta_L]_D$$
$$= c_1 \eta^{\mu\nu} \phi^*_{,\mu}(x)\phi_{,\nu}(x) = c_1 \partial_\mu \phi^*(x).\partial^\mu \phi(x)$$
$$T_2 = \phi^*(x)\phi_{,\mu\nu}(x)[\theta_R^T \epsilon\gamma^\mu \theta_L.\theta_R^T \epsilon\gamma^\nu \theta_L]_D$$

$$= c_1\phi^*(x)\Box\phi(x)$$

where

$$\Box = \eta^{\mu\nu}\partial_\mu\partial_\nu = \partial^\mu\partial_\mu$$

and likewise,

$$T_3 = c_1\phi(x)\Box\phi^*(x)$$

Remark:

$$\theta_R^T\epsilon\gamma^\mu\theta_L.\theta_R^T\epsilon\gamma^\nu\theta_L$$

$$= \theta_R^T\epsilon\gamma^\mu\theta_L\theta_L^T\epsilon\gamma^\nu\theta_R$$

$$\theta_L\theta_L^T = ((1+\gamma^5)/2)\theta\theta^T((1+\gamma^5)/2)$$

$$= \theta\epsilon(1+\gamma^5)\theta((1+\gamma^5)/2)\epsilon$$

$$= \theta_L^T\epsilon\theta_L(1+\gamma^5)\epsilon$$

Thus,

$$\theta_R^T\epsilon\gamma^\mu\theta_L.\theta_R^T\epsilon\gamma^\nu\theta_L =$$

$$(\theta_L^T\epsilon\theta_L)\theta_R^T\epsilon\gamma^\mu\epsilon(1+\gamma^5)\epsilon\epsilon\gamma^\nu\theta_R$$

$$= -2(\theta_L^T\epsilon\theta_L).\theta_R^T\epsilon\gamma^\mu\epsilon\gamma^\nu\theta_R$$

$$= c_1(\theta^T\epsilon\theta)^2\eta^{\mu\nu}$$

which follows on expressing γ^μ in terms of the Pauli spin matrices and using the fact that θ_L has components $\theta_{1:2}$ while θ_R has components $\theta_{3:4}$ and the product of any four of the $\theta's$ is non-zero iff all the $\theta's$ are distinct. Again,

$$[\theta_R^T\gamma^0\psi^*(x_-)\theta_L^T\epsilon\psi(x_+)]_D$$

$$= [\psi_{,\mu}^*(x)^T\theta_R^T\epsilon\gamma^\mu\theta_L\gamma^0\theta_R\theta_L^T\epsilon\psi(x)]_D$$

$$+[\psi_{,\mu}(x)^T\theta_R^T\epsilon\gamma^\mu\theta_L\epsilon\theta_L\theta_R^T\gamma^0\psi^*(x)]_D$$

Now,

$$\theta_R\theta_L^T = \theta_R^T\epsilon\gamma^\mu\theta_L\epsilon\gamma_\mu(1+\gamma^5)$$

So

$$[\psi_{,\mu}^*(x)^T\theta_R^T\epsilon\gamma^\mu\theta_L\gamma^0\theta_R\theta_L^T\epsilon\psi(x)]_D$$

$$= 2[\theta_R^T\epsilon\gamma^\mu\theta_L.\theta_R^T\epsilon\gamma^\nu\theta_L]_D.\psi_{,\mu}^*(x)^T\gamma^0\epsilon\gamma_\nu\epsilon\psi(x)$$

$$= -2c_1\eta^{\mu\nu}\psi_{,\mu}^*(x)^T\gamma^0\gamma_\nu^T\psi(x)$$

$$= -2c_1\eta^{\mu\nu}\bar{\psi}_{,\mu}(x)\gamma_\nu^T\psi(x)$$

$$= -2c_1\eta^{\mu\nu}\psi_{,\mu}^*(x)^T\gamma_\nu^{*T}\gamma^0\psi(x)$$

$$= -2c_1(\gamma^\mu\psi_{,\mu}(x))^*\gamma^0\psi(x)$$

where we have used the fact that $\gamma_\mu\gamma^0$ is Hermitian and hence its transpose $\gamma^0\gamma_\mu^T$ is also Hermitian. Note that the conjugate of this quantity is given by

$$-2c_1\psi(x)^*\gamma^0\gamma^\mu\psi_{,\mu}(x) = -2c_1\bar{\psi}(x)\gamma^\mu\psi_{,\mu}(x)$$

$$= -2c_1\psi(x)^*\alpha^\mu\psi_{,\mu}(x)$$

Supercurrent:
Feynnman superpath integrals and superpropagators.
The action functional for the left Chiral superfield Φ is taken as

$$S[\Phi] = \int [\Phi^*\Phi]_D d^4x + \int [f(\Phi)]_F d^4x$$

Equivalently, writing $\Phi = D_R^2 S$ we can express this as

$$S = \int (D_L^2 S^*)(D_R^2 S)d^4xd^4\theta + \int \tilde{S}d^4xd^4\theta$$

Note that

$$f(D_R^2 S) = D_R^2\tilde{S}$$

for some other superfield $\tilde{S}$. For example,

$$(D_R^2 S)^2 = D_R^2(SD_R^2 S), (D_R^2 S)^3 = D_R^2(S(D_R^2 S)^2)$$

etc. So in fact writing

$$f(\Phi) = \sum_{k\geq 1} c(k)\Phi^k$$

we get

$$f(D_R^2 S) = D_R^2(\sum_{k\geq 1} c(k)S(D_R^2 S)^{k-1})$$

So formally, we can write

$$f(D_R^2 S) = D_R^2(S.(D_R^2 S)^{-1}f(D_R^2 S))$$

The superfield equations are expressible as

$$D_R^2 D_L^2 S^* = f'(\Phi) = f'(D_R^2 S)$$

or equivalently,

$$D_L^2 D_R^2 S = f*'(D_L^2 S^*)$$

since

$$f'(D_R^2 S)D_R^2\delta S = D_R^2(f'(D_R^2 S)\delta S)$$

and hence

$$\int [f'(D_R^2 S)D_R^2\delta S]_F d^4x = \int f'(D_R^2 S)\delta Sd^4xd^4\theta$$

By analogy with quantum field theory, it is therefore natural to consider a super-Green's function $G(x,\theta|x',\theta')$ that satisfies the propagator equation

$$D_L^2 D_R^2 G(x,\theta|x',\theta') - f'*(D_L^2 G(x,\theta|x',\theta')^*) = \delta^4(x-x')\delta^4(\theta-\theta')$$

Note that by the definition of the Berezin/Fermionic integral,

$$\int \theta_{i_1}..\theta_{i_k} d^4\theta$$

is zero if either $k < 4$ or if $k > 4$ while

$$\int \theta_1\theta_2\theta_3\theta_4 d^4\theta = 1$$

it follows that

$$\delta^4(\theta - \theta') = (\theta_1 - \theta_1')(\theta_2 - \theta_2')(\theta_3 - \theta_3')(\theta_4 - \theta_4')$$

This may be explicitly checked by writing out $f(\theta)$ as

$$f(\theta) = c_0 + c_1(k)\theta_k + c_2(k,m)\theta_k\theta_m + c_3(k,m,n)\theta_k\theta_m\theta_n + c_4\theta_1\theta_2\theta_3\theta_4$$

and applying the above Berezin rules taking into account the anitcommutativity of the θ and the θ' to show that

$$\int f(\theta)\delta^4(\theta - \theta')d^4\theta = f(\theta')$$

In the absence of a superpotential $f(\Phi)$, the field equations are

$$D_L^2 D_R^2 S = 0$$

This equation should be regarded as the super-version of the classical massless Klein-Gordon equation or equivalently the wave equation. The corresponding super-propagator G should satisfy the super pde

$$D_L^2 D_R^2 G = P\delta^4(x - x')\delta^4(\theta - \theta')$$

where P is the projection onto the space of superfields fields that belong to the orthogonal complement of the nullspace of $D_L^2 D_R^2$, or equivalently that belong to the range space of $D_L^2 D_R^2$. This is precisely in analogy with quantum electrodynamics. It is easy to see that

$$P = K\Box^{-1} D_L^2 D_R^2$$

for some real constant K. In fact, we have that P annihilates any vector in the range of D_R and hence in the range of D_R^2 and further, we have

$$P^2 = K^2\Box^{-2} D_L^2 D_R^2 D_L^2 D_R^2 = D_L^2[D_R^2, D_L^2]D_R^2$$

with

$$D_R^2 D_{La} = D_R^T \epsilon D_R D_{La}$$

$$\{D_{Ra}, D_{Lb}\} = \{(\gamma^\mu \theta_L \partial_\mu - \gamma^5 \epsilon \partial_{\theta_R})_a, (\gamma^\nu \theta_R \partial_\nu - \gamma^5 \epsilon \partial_{\theta_L})_b\}$$

$$= -(\gamma^\mu)_{ac}\{\theta_{Lc}, \partial_{\theta_{Ld}}\}(\gamma^5\epsilon)_{bd}\partial_\mu$$
$$-(\gamma^5\epsilon)_{ac}\{\partial_{\theta_{Rc}}, \theta_{Rd}\}\gamma^\nu_{bd}\partial_\nu$$
$$= -[\gamma^\mu((1+\gamma^5)/2)(\gamma^5\epsilon)^T - \gamma^5\epsilon((1-\gamma^5)/2)\gamma^{\mu T}]_{ab}\partial_\mu$$
$$= [\gamma^\mu(1+\gamma^5)\epsilon + (1-\gamma^5)\gamma^\mu\epsilon]_{ab}\partial_\mu$$
$$= [\gamma^\mu\epsilon(1+\gamma^5)]_{ab}\partial_\mu = X_{ab}$$

say. Interchanging a and b gives

$$\{D_{La}, D_{Rb}\} = -((1+\gamma^5)\epsilon\gamma^{\mu T})_{ab}\partial_\mu$$
$$= -(\gamma^\mu\epsilon(1-\gamma^5))_{ab}\partial_\mu = X_{ba}$$

say. In matrix notation, these identities are expressible as

$$\{D_R, D_L^T\} = \gamma^\mu\epsilon(1+\gamma^5)\partial_\mu,$$
$$\{D_L, D_R^T\} = -\gamma^\mu\epsilon(1-\gamma^5)\partial_\mu$$

Adding these two equations and noting that D_L anticommutes with itself and D_R also anticommutes with itself, we get

$$\{D, D^T\} = 2\gamma^\mu\gamma^5\epsilon\partial_\mu$$

Then,

$$D_R^2 D_{La} = \epsilon_{bc}D_{Rb}D_{Rc}D_{La} =$$
$$\epsilon_{bc}D_{Rb}(X_{ca} - D_{La}D_{Rc})$$
$$= \epsilon_{bc}X_{ca}D_{Rb} - \epsilon_{bc}D_{Rb}D_{La}D_{Rc}$$
$$= \epsilon_{bc}X_{ca}D_{Rb} - \epsilon_{bc}(X_{ba} - D_{La}D_{Rb})D_{Rc}$$
$$= \epsilon_{bc}X_{ca}D_{Rb} - \epsilon_{bc}X_{ba}D_{Rc}$$
$$+D_{La}D_R^2$$

Equivalently,

$$[D_R^2, D_{La}] = \epsilon_{bc}X_{ca}D_{Rb} - \epsilon_{bc}X_{ba}D_{Rc}$$

Then,

$$D_R^2 D_{La}D_{Lp} = \epsilon_{bc}X_{ca}(X_{bp} - D_{Lp}D_{Rb})$$
$$-\epsilon_{bc}X_{ba}(X_{cp} - D_{Lp}D_{Rc}) + D_{La}([D_R^2, D_{Lp}] + D_{Lp}D_R^2)$$

so that

$$D_R^2 D_L^2 D_R^2 = \epsilon_{ap}D_R^2 D_{La}D_{Lp}D_R^2 =$$
$$= (epsilon_{bc}X_{ca}\epsilon_{ap}X_{bp} - \epsilon_{bc}X_{ba}\epsilon_{ap}X_{cp})D_R^2$$

since the product of any three $D'_R s$ is zero. We can express this relationship as

$$(D_R^2 D_L^2)^2 =$$

$$2Tr(\epsilon.X.\epsilon.X^T)D_R^2$$

and hence,

$$(D_L^2 D_R^2)^2 = 2Tr(\epsilon.X.\epsilon.X^T)D_L^2 D_R^2$$

Now,

$$X = \gamma^\mu \epsilon(1+\gamma^5)\partial_\mu$$

and hence,

$$Tr(\epsilon.X.\epsilon.X^T) =$$

$$-Tr(\epsilon.\gamma^\mu \epsilon(1+\gamma^5)^2 \epsilon^2 \gamma^{\nu T})\partial_\mu \partial_\nu$$

But,

$$-Tr(\epsilon.\gamma^\mu \epsilon(1+\gamma^5)^2 \epsilon^2 \gamma^{\nu T})$$

$$2Tr(\epsilon.\gamma^\mu.\epsilon(1+\gamma^5)\gamma^{\nu T})$$

$$= -2Tr(\epsilon \gamma^\mu \gamma^\nu (1-\gamma^5)\epsilon)$$

$$= 2.Tr(\gamma^\mu \gamma^\nu (1-\gamma^5)) = c.\eta^{\mu\nu}$$

where c is a real constant. Thus,

$$(D_L^2 D_R^2)^2 = c.\Box.D_L^2 D_R^2$$

from which it follows that $P = c^{-1} D_L^2 D_R^2/\Box$ is idempotent, ie, a projection.

Supersymmetric gauge theories: The gauge

Design of quantum unitary gates using supersymmetric field theories:

Given a Lagrangian for a set of Chiral superfields and gauge superfields, we can construct the action as an integral of the Lagrangian over space-time. We can include forcing terms in this Lagrangian for example by adding c-number control gauge potentials to the quantum gauge field $V_\mu^A(x)$ or c-number control current terms to the terms involving the Dirac current which couples to the gauge field. After adding these c-number control terms, the resulting action will no longer be supersymmetric. However, we can still construct the Feynman path integral for the resulting action between two states of the field ie, by specifying the fields at the two endpoints of a time interval $[0, T]$ and then we obtain a transition matrix between these two states of the field. For example, the initial state can be a coherent state in which the annihilation component of the electromagnetic vector potential has definite values and the Dirac field of electrons and positrons is in a Fermionic coherent state where the annihilation component of the wave function has definite values. Likewise with the final state. Or else, we may specify the initial state to be a state in which there are definite numbers of photons, electrons and positrons having definite four momenta and spins and so also with the final state. In the case of a supersymmetric theory, we'll have to also specify the states of the other fields like the gaugino field, the gravitino field and the auxiliary fields or else we may break the supersymmetry by expressing the auxiliary fields in terms of the other superfield components using the variational equations of motion and then calculate the the Feynman path integral corresponding to an initial and a final state and then make this transition matrix as close as possible to a desired transition matrix by optimizing over the c-number control fields.

[21] One of the main achievements in the work of C.R.Rao was the proof of the lower bound on the error covariance matrix of a statistical estimator of a vector valued parameter based on vector valued observations using techniques of matrix theory. C.R.Rao in his work has also considered the case when the Fisher information matrix is singular and in this case, he has been able to use the methods of generalized inverses to obtain new formulas for the lower bound. The lower bound on the variance of an estimator should be compared to the Heisenberg uncertainty principle in quantum mechanics for two non-commuting observables. In fact, it can be shown that the Heisenberg uncertainty inequality for position and momentum can be derived using the CRLB. The CRLB roughly tells us that no matter how much we may try, we can never achieve complete accuracy in our estimation process, ie, there is inherently some amount of uncertainty about the system that generates a random observation.

Chapter 3

Some Study Projects on Applied Signal Processing with Remarks About Related Contributions of Scientists

[1] Linear models: Time series models like $AR, MA, ARMA$, casting these models in the form

$$X(n) = H(n)\theta + V(n)$$

where $X(n), H(n)$ are data vectors and data matrices. $V(n)$ is noise. $H(n) \in \mathbb{R}^{n \times p}, X(n), V(n) \in \mathbb{R}^n$. If $R_v = Cov(V(n))$ and $V(n)$ are iid zero mean Gaussian, then the MLE of θ based on data collected upto time n is given by

$$\hat{\theta}(n) = (\sum_{k=1}^{n} H(k)^T R_v H(k))^{-1} (\sum_{k=1}^{n} H(n)^T R_v X(k))$$

Since $X(n), H(n)$ are data matrices, they are also random and we wish to determine the mean and covariance of $\hat{\theta}(n)$ in terms of the statistics of these data matrices.

[2] Innovations process and its application to the construction of the Wiener filter. Stochastic processes as curves in Hilbert space. Let $x(n)$ be a stationary process. Its innovations process $e(n)$ can be expressed as

$$e(n) = P^{\perp}_{span(x(k):k \leq n-1)} x(n) = x(n) - P_{span(x(k).k \leq n)} x(n)$$

$$= \sum_{k=0}^{\infty} l(k) x(n-k), l(0) = 1$$

This means that

$$e(n) \in span(x(k) : k \leq n)$$

and by inversion,

$$x(n) \in span(e(k) : k \leq n)$$

so we can write

$$L(z) = \sum_{k \geq 0} l(k) z^{-k}, H(z) = L(z)^{-1} = \sum_{k \geq 0} h(k) z^{-k}, x(n) = \sum_{k \geq 0} h(k) e(n-k), h(0) = 1$$

$e(n)$ is clearly a white process. Let

$$\mathbb{E}(e(n)^2) = \sigma_e^2$$

Then, the power spectral density of $x(n)$ is given by

$$S_{xx}(z) = \sigma_e^2 H(z) H(z^{-1})$$

To get a causal, stable $H(z)$ with an inverse that is also causal and stable, we assume that $S_{xx}(z)$ is rational and then select $H(z)$ so that it is causal stable and minimum phase, ie, is poles and zeroes all fall within the unit circle. Then $L(z) = H(z)^{-1}$ is also obviously causal and minimum phase. The prediction error energy is

$$\sigma_e^2 = h(0)^2$$

and if Γ denotes the unit circle,

$$(2\pi)^{-1} \int_\Gamma ln(S_{xx}(z)) z^{-1} dz = \sigma_e^2$$

This can be easily verified using the residue theorem and the fact that $H(z)$ has all its poles and zeroes inside the unit circle.

[3] **Some Study Project problems**

[1] Compute the capacitance between two parallel cylinders having different radii.

[2] State and prove Doob's maximal inequality for submartingales.

[3] State and prove the Martingale down-crossing inequality and its application to proving the martingale convergence theorem.

[4] State and prove Doob's L^p-inequality for martingales.

[5] Power spectrum estimation:Compute the mean and covariance of the periodogram of a stationary Gaussian process.

[6] Apply of Doob's L^2-Martingale inequality to prove the almost sure existence and uniqueness of the solutions to Ito's stochastic differential equation when the drift and diffusion coefficients satisfy the Lipshitz conditions.

[7] When the Choi-Kraus-Stinespring operators of a quantum noisy channel have classical randomness, then how does one determine the mean square state estimation error of the output of the recovery channel when the recovery operators have been designed in accordance with the Knill-Laflamme theorem for

the mean value of the noisy channel operators in the Choi-Kraus-Stinespring representation ?

[8] Derive the Nonlinear filtering equations for a Markov state when the measurement noise is a mixture of a white Gaussian component and a compound Poisson component.

[9] In quantum scattering theory, when the free particle Hamiltonian is H_0 and the scattering potential is V where V is a Gaussian random Hermitian operator, then how does one compute the statistical moments of the scattering operator using the well known formula for the moments of a Gaussian random vector.

[10] Application of Cramer's theorem to computing the optimal rate at which the probability of missing the target tends to zero given that the probability of false alarm tends to zero with regard to the binary hypothesis testing problem for a sequence of iid random variables, ie, under H_1, the data $(X_1, ..., X_n)$ has a pdf of $p_1(x_1)...p_1(x_n)$ and under H_0, it has a pdf of $p_0(x_1)...p_0(x_n)$. The Neyman-Pearson test for n iid data samples is given as follows: Select H_1 if $\frac{p_1(x_1)...p_1(x_n)}{p_0(x_1)...p_0(x_n)} > \lambda_n$ and select H_0 otherwise, where λ_n is chosen so that $Pr(H_1|H_0) = P_F(n) \to 0$ as $n \to \infty$ and we prove that under this constraint on $P_F(n)$, the minimum possible value of $lim_n n^{-1} log(P_M(n))$ is $-D(p_1|p_0)$ where $P_M(n) = Pr(H_0|H_1)$ and

$$D(p_1|p_0) = \int p_1(x).ln(p_1(x)/p_0(x))dx$$

The proof of this result is based on defining sequence

$$\xi(n) = log(p_1(X_n)/p_0(X_n))$$

and

$$S(n) = (\xi(1) + ... + \xi(n))/n$$

Under any given hypothesis, the $\xi(n)'s$ are iid r.v's. The Neyman-Pearson test is $S(n)/n > \eta(n)$ implies select H_1 and select H_0 otherwise where $\eta(n) = n^{-1}.log(\lambda_n)$. The false alarm probability is given by

$$P_F(n) = Pr(S(n)/n > \eta(n)|H_0)$$

and the miss probability is

$$P_M(n) = Pr(S(n)/n < \eta(n)|H_1)$$

These two probabilities are approximately evaluated using Cramer's theorem. We have

$$n^{-1} log(\mathbb{E}(exp(s.S_n)|H_0) = log\mathbb{E}(exp(s.X_1)|H_0)$$

$$= log \int (p_1(x)/p_0(x))^s p_0(x)dx = log \int p_1(x)^s p_0(x)^{1-s} dx$$

and

$$n^{-1} log(\mathbb{E}(exp(s.S_n)|H_1) = log \int p_1(x)^{1+s} p_0(x)^s dx$$

Thus, for large n,

$$n^{-1}.log(P_F(n)) \approx -inf_{x>\eta}sup_s(s.x - log\int p_1(x)^s p_0(x)^{1-s}dx)$$

$$= -sup_{s\geq 0}(s.\eta - log\int p_1(x)^s p_0(x)^{1-s}dx)$$

The minimum that we require is that $P_F(n) \to 0$ and this is guaranteed once the above quantity is negative, in the worst case, we may take it to be a negative number arbitrarily close to zero. Equivalently, in this worst case situation, the supremum above, namely zero is attained when $s = 0$ and the optimal choice of the threshold η is obtained by setting the derivative above w.r.t s to be zero at $s = 0$, ie,

$$\eta = \int p_1(x)log(p_1(x)/p_0(x))dx = D(p_1|p_0)$$

For this value of the optimal threshold, we compute the optimal rate at which $P_M(n)$ converges to zero again by applying Cramer's theorem.

[11] Basics of queueing theory:

Let $X_1, X_2, ...$ denote the successive interarrival times of packets in a single server queue and let $T_1, T_2, ...$ denote the service times for packet 1, packet 2,...etc. Let W_n denote the total waiting time for the nth packet including the service time, ie, W_n is the time taken for the nth packet starting from his arrival time upto the time when his service is completed and he leaves. We then have the obvious recursive relationship

$$W_{n+1} = max(S_n + W_n - S_{n+1}, 0) + T_{n+1}$$

where

$$S_n = X_1 + ... + X_n$$

Suppose the $X_n's$ are iid with distribution F_X and the $T_n's$ are iid with distribution F_T. Then, the probability is to determine the law of the waiting time process $\{W_n, n = 1, 2, ...\}$. We also wish to determine the distribution of $N(t)$, the number of packets in the queue at time t. We see that the total number of departures that have taken place in the duration $[0, t]$ is given by $D(t) = max(n \geq 1 : W_1 + ... + W_n \leq t\}$ and the total number of arrivals that have taken place in the duration $[0, t]$ is given by $A(t) = max\{n \geq 1 : S_n \leq t\}$. Then, the size of the queue at time t is given by

$$N(t) = A(t) - D(t)$$

[12] Group representation theory and its application to statistical image processing on a curved manifold.

[1] Definition of the image field model on a manifold $\mathcal{M}$ on which a Lie group of transformations G acts.

[2] Estimation of the group transformation element from the measured image field with knowledge of the original noiseless untransformed image field using the irreducible representations of the group G.

[3] by assuming that the estimate of the G-transformation element g is a small perturbation of the true transformation, ie,

$$g = exp(\delta.X)g_0, X \in \mathfrak{g}$$

calculate the value of X upto $O(\delta^m)$ and hence determine the probability that the error X is larger than a threshold ϵ for a given g_0.

Some details:The image field model

$$f(x) = f_0(g_0^{-1}x) + w(x), x \in \mathcal{M}$$

Let $Y_{nl}(x), l = 1, 2, ..., d_n$ define an onb for an irreducible unitary representation π_n of G appearing in the decomposition of the representation U in $L^2(\mathcal{M}, \mu)$ where μ is a G-invariant measure on $\mathcal{M}$ and $U(g)f(x) = f(g^{-1}.x)$. Then

$$U(g)Y_{nl}(x) = Y_{nl}(g^{-1}x) = \sum_{m=1}^{d_n} [\pi_n(g)]_{ml} Y_{nm}(x)$$

with the additional condition,

$$\int_{\mathcal{M}} \bar{Y}_{nl}(x)Y_{km}(x)d\mu(x) = \delta(n, k).\delta(l, m)$$

and

$$L^2(\mathcal{M}, \mu) = Cl(span\{Y_{nl} : 1 \leq l \leq d_n, n \geq 1\})$$

The image field model is then equivalent to

$$\mathbf{f}(n) = \pi_n(g)\mathbf{f}_0(n) + \mathbf{w}(n), n \geq 1$$

where

$$\mathbf{f}(n) = ((f(n, l)))_{l=1}^{d_n}, f(n, l) = \int_{\mathcal{M}} f(x)\bar{Y}_{nl}(x)d\mu(x)$$

Let g_0 be the true value of g and $exp(\delta.\hat{X})g_0$ its estimate. Then, we have

$$\hat{X} = argmin_{X \in \mathfrak{g}} \sum_n \sigma(n)^{-2} \parallel \mathbf{f}(n) - \pi_n(exp(\delta.X)g_0)\mathbf{f}_0(n) \parallel^2$$

where we are assuming that the noise is Gaussian with zero mean and G-invariant correlation.

[13] Statistical theory of fluid turbulence:Partial differential equations for the velocity field moments. From homogeneity and isotropy,

$$R_{ij}(r_1, r_2) = \langle v_i(r_1)v_j(r_2) \rangle = A(r)n_i n_j + B(r)\delta_{ij}$$

where $r = |r_1 - r_2|$ and $\hat{n} = (r_2 - r_1)/|r_2 - r_1|$. Also

$$C_{ijk}(r_1, r_2, r_3) =< v_i(r_1)v_j(r_2)v_k(r_3) >= C_{ijk}(r_2 - r_1, r_3 - r_1)$$

This third rank tensor must be constructed using scalar functions of $|r_2 - r_1|, |r_3 - r_1|$ and the unit vectors along the directions $r_2 - r_1$ and $r_3 - r_1$ and it must be symmetric w.r.t the interchange of $(r_2 - r_1, j)$ and $(r_3 - r_1, k)$. Thus the general form of this tensor is given by

$$C_{ijk}(r_1, r_2, r_3) = A_1(|r_2 - r_1|, |r_3 - r_1|)n_i n_j m_k + A_1(|r_3 - r_1|, |r_2 - r_1|)m_i m_j n_k$$

$$A_3(|r_2 - r_1|, |r_3 - r_1|)\delta_{ij} m_k + A_3(|r_3 - r_1|, |r_2 - r_1|)\delta_{ik} n_j$$

$$+A_4(|r_2 - r_1|, |r_3 - r_1|)n_i n_j n_k + A_4(|r_3 - r_1|, |r_2 - r_1|)m_i m_j m_k$$

where n is the unit vector along $r_2 - r_1$ and m is the unit vector along $r_3 - r_1$.

[4] **Estimating the parameters in an ARMA model**

$$X_N = H_N a + G_N b$$

where

$$X_N = [x(N), x(N-1), ..., x(0)]^T, W_N[w(N), w(N-1), ..., w(0)]^T$$

$$H_N = [z^{-1}X_N, ..., z^{-p}X_N], G_N = [W_N, z^{-1}W_N, ..., z^{-q}W_N]$$

$$a = [a(1), ..., a(p)]^T, b = [b(0), ..., b(q)]^T$$

This defines the ARMA model. In order to estimate a, b from this model, we require to compute the pdf of X_N given a, b and then maximize this pdf over a, b.

[5] **Statistical properties of parameter estimates in the AR model using matrix perturbation theory**

$$X_N = H_N\theta + W_N$$

where

$$X_N = [x(N), x(N-1), ..., x(0)]^T,$$

$$H_N = [z^{-1}X_N, z^{-2}X_N, ..., z^{-p}X_N]$$

$$W_N = [w(N), w(N-1), ..., w(0)]^T$$

$w(n)'s$ are assumed to be iid $N(0, \sigma^2)$.

$$\hat{\theta}(N) = (H_N^T H_N)^{-1} H_N^T X_N$$

Now,

$$N^{-1}H_N^T H_N = ((N^{-1}z^{-i}X_N^T z^{-j}X_N))_{1\le i,j\le p} = \hat{R}$$

$$N^{-1}H_N^T X_N = ((N^{-1}z^{-i}X_N^T X_N))_{i=1}^p = \hat{r}$$

where

$$r = ((r(i))), r(i) = R(i) = \mathbb{E}(x(n-i)x(n)), R = ((R(i-j)))_{1 \le i,j \le p}$$

$$\hat{r}(i) = N^{-1} \sum_{n=1}^{N} x(n-i)x(n), \hat{R}(i,j) = N^{-1} \sum_{n=1}^{N} x(n-i)x(n-j)$$

We write

$$\hat{R} = R + \delta R, \hat{r} = r + \delta r$$

Clearly,

$$\theta = R^{-1} r$$

$$\hat{\theta}(N) = \hat{R}^{-1}\hat{r} = (R + \delta R)^{-1}(r + \delta r)$$

$$\approx (R^{-1} - R^{-1}\delta R.R^{-1})(r + \delta r)$$

$$= \theta + R^{-1}\delta r - R^{-1}\delta R.\theta$$

so the estimation error is given by

$$e_N = \hat{\theta}(N) - \theta = R^{-1}\delta r - R^{-1}\delta R.\theta$$

Large deviation evaluation of the rate at which e_N converges to zero as $N \to \infty$. Note that by ergodicity,

$$\delta r, \delta R \to 0, N \to \infty$$

Evaluation of the rate at which these converges to zero amounts to evaluating the rate at which $z[n] = N^{-1} \sum_{n=1}^{N} y(n)$ converges to zero for any stationary process $y(n)$. We make use of the Gartner-Ellis theorem to evaluate the LDP rate:

$$\Lambda_N(\lambda) = N^{-1}.log(\mathbb{E}exp(N\lambda.z[N]))$$

$$= N^{-1}.log(\mathbb{E}exp(\lambda. \sum_{n=1}^{N} y(n)))$$

If $y(n)$ is approximately Gaussian with zero mean with autocorrelation $R(n)$, then the above equals

$$N^{-1}.log(exp((\lambda^2/2) \sum_{|n| \le N-1} (N-1-|n|)R(n)))$$

$$= (N^{-1}\lambda^2/2) \sum_{|n| \le N-1} (N-1-|n|)R(n)$$

which converges as $N \to \infty$ to

$$(\lambda^2/2) \sum_{n=-\infty}^{\infty} R(n) = \lambda^2 S(0)/2$$

where

$$S(\omega) = \sum_n R(n)exp(-j\omega n)$$

is the power spectral density of $y(n)$. This determines the large deviation rate. It is more complicated to derive a formula for the rate at which the empirical distribution $L_N(.) = N^{-1}\sum_{n=1}^N \delta_{y(n)}$ converges to the one dimensional marginal distribution of $y(n)$. For this, we require to evaluate the limiting Gartner-Ellis logarithmic moment generating function

$$\Lambda_f = lim_{N\to\infty} N^{-1}.log(\mathbb{E}(exp(\sum_{n=1}^N f(y(n))))$$

[6] Proof of the L^2-mean ergodic theorem for wide sense stationary processes under the condition $C(k) \to 0, |k| \to \infty$.

$$C(k) = R(k) - \mu^2, \mu = \mathbb{E}(x(n)), R(k) = \mathbb{E}(x(n)x(n+k))$$

Let

$$S_N = \sum_{n=1}^N x(n)$$

$$\mathbb{E}(S_N/N - \mu)^2 = N^{-1} \sum_{|k|\leq N-1} (1-(1+|k|)/N)C(k) \to 0, N \to 0$$

provided that

$$C(N) \to 0$$

This follows by the Cesaro theorem: If $a_n \to 0, n \to 0$, then $n^{-1}\sum_{k=1}^n a_k \to 0$. This proves the mean ergodic theorem for wide sense stationary processes. Now let $x(n)$ be a stationary Gaussian process. Then fix $k \in \mathbb{Z}$ and put $y(n) = (x(n) - \mu_x)(x(n+k) - \mu_x)$. Then $y(n)$ is also a stationary process with

$$\mathbb{E}(y(n)) = C_x(k), C_y(m) = \mathbb{E}(y(n+m)y(n)) - C_x(k)^2 =$$

$$= \mathbb{E}[(y(n+m) - C_x(k))(y(n) - C_x(k))]$$

$$= C_x(m)^2 + C_x(m+k)C_x(m-k) \to 0, |m| \to \infty$$

provided that $C_x(m) \to 0, |m| \to \infty$. Thus, by the mean ergodic theorem applied to $y(n)$,

$$N^{-1}\sum_{n=1}^N y(n) \to C_x(k)$$

which is the same as saying that

$$N^{-1}.\sum_{n=1}^N x(n)x(n+k) \to R_x(k)$$

since

$$N^{-1}\sum_{n=1}^{N} x(n) \to \mu_x$$

by the mean ergodic theorem applied to $x(n)$.

[7] **Quantum filtering of cavity resonator fields in interaction with a bath.** Consider first the TM modes:

$$H_z = 0, E_z(t,x,y,z) = \sum_{mnp} Re(c(mnp)exp(j\omega(mnp)t))u_{mnp}(x,y,z)$$

$$E_\perp(t,x,y,z) = \sum_{mnp} h_{mn}^{-2} Re(c(mnp)exp(j\omega(mnp)t))\partial_z \nabla_\perp u_{mnp}(x,y,z)$$

or equivalently,

$$E_x(t,x,y,z) = \sum_{mnp} h_{mn}^{-2}(m\pi/a)(p\pi/d)Re(c(mnp)exp(j\omega(mnp)t))v_{mnp}(x,y,z)$$

$$E_y(t,x,y,z) = \sum_{mnp} h_{mn}^{-2}(n\pi/b)(p\pi/d)Re(c(mnp)exp(j\omega(mnp)t))w_{mnp}(x,y,z)$$

where

$$u_{mnp}(x,y,z) = ((2.\sqrt{2})/\sqrt{abd})sin(m\pi x/a)sin(n\pi y/b)cos(p\pi z/d)$$

$$v_{mnp}(x,y,z) = -((2.\sqrt{2})/\sqrt{abd})cos(m\pi x/a)sin(n\pi y/b)sin(p\pi z/d)$$

$$w_{mnp}(x,y,z) = -((2.\sqrt{2})/\sqrt{abd})sin(m\pi x/a)cos(n\pi y/b)sin(p\pi z/d)$$

Let $< . >$ denote time average. Then

$$\int_{box} < E_z^2 > dxdydz = (1/2)\sum_{mnp} |c(mnp)|^2,$$

$$\int_{box} < E_x^2 + E_y^2 > dxdydz = (1/2)\sum_{mnp} |c(mnp)|^2 h_{mn}^{-4}(\pi p/d)^2(h_{mn}^2)$$

$$= (1/2)\sum_{mnp} |c(mnp)|^2((\pi p/d)^2/h_{mn}^2)$$

where

$$h_{mn}^2 = \pi^2(m^2/a^2 + n^2/b^2)$$

The total energy in the cavity due to the electric field is then

$$(\epsilon/2)\int_{box}(E_z^2 + |E_\perp^2)dxdydz$$

$$= \sum_{mnp} \lambda(mnp)|c(mnp)|^2$$

which can be abbreviated to

$$H_s = \sum_n \omega(n)c(n)^*c(n)$$

where $c(n)$ are annihilation operators of the cavity field and $c(n)^*$ are the creation operators. They satisfy the Boson CCR:

$$[c(n), c(m)^*] = \delta[n-m]$$

Bath field is

$$E_B(t,r) = \sum_k [A'_k(t)\psi_k(r) + A'_k(t)^*\bar{\psi}_k(r) + \Lambda'_k(t)\eta_k(r)]$$

where $A_k(.), A_k(.)^*, \Lambda_k(.)$ are the fundamental noise processes in the quantum stochastic calculus of Hudson and Parthasarathy. They satisfy the quantum Ito formula

$$dA_k dA_m^* = \delta_{k,m}dt, d\Lambda_k.d\Lambda_m = \delta_{km}d\Lambda_k,$$

$$dA_k d\Lambda_m = \delta_{km}dA_k, d\Lambda_m dA_k^* = \delta_{mk}dA_k^*$$

Denote the above system electric field E by $E_s(t,r)$. Then the total field energy of the system plus bath fields within the cavity resonator is given by

$$(\epsilon/2)\int_{box} |E_s(t,r) + E_B(t,r)|^2 d^3r$$

Ignoring the bath energy, the total field energy of the system (ie, cavity resonator) plus its interaction energy with the bath is given by

$$H(t) = H_s + H_I(t)$$

where

$$H_s = (\epsilon/2)\int_{box} |E_s(t,r)|^2 d^3r,$$

$$H_I(t) = \epsilon\int_{box} (E_s(t,r), E_B(t,r))d^3r$$

$$= \sum_k (L_k(t)dA_k + M_k(t)dA_k^* + N_k(t)d\Lambda_k)$$

with the $L_k(t), M_k(t), N_k(t)$ being system operators defined by

$$L_k(t) = \int_{box} \psi_k(r)E_s(t,r)d^3r, M_k(t) = \int_{box} \psi_k(r)^*E_s(t,r)d^3r,$$

$$N_k(t) = \int_{box} \eta_k(r)E_s(t,r)d^3r$$

Writing

$$E_s(t,r) = \sum_n Re(c(n).exp(j\omega(n)t))F_n(r)$$

we get

$$L_k(t) = \sum_n (l_{1kn}(t)c(n) + l_{2kn}(t)c(n)^*)$$

where

$$l_{1kn}(t) = (1/2)exp(j\omega(n)t)\int_{box} \psi_k(r)F_n(r)d^3r$$

$$l_{2kn}(t) = (1/2)exp(-j\omega(n)t)\int_{box} \psi_k(r)F_n(r)d^3r$$

$$M_k(t) = \sum_n (m_{1kn}(t)c(n) + m_{2kn}(t)c(n)^*)$$

where

$$m_{1kn}(t) = (1/2)exp(j\omega(n)t)\int_{box} \bar{\psi}_k(r)F_n(r)d^3r$$

$$m_{2kn}(t) = (1/2)exp(-j\omega(n)t)\int_{box} \bar{\psi}_k(r)F_n(r)d^3r$$

and

$$N_k(t) = \sum_n (n_{1kn}(t)c(n) + n_{2kn}(t)c(n)^*)$$

where

$$n_{1kn}(t) = (1/2)exp(j\omega(n)t)\int_{box} \eta_k(r)F_n(r)d^3r$$

$$n_{2kn}(t) = (1/2)exp(-j\omega(n)t)\int_{box} \eta_k(r)F_n(r)d^3r = \bar{n}_{1kn}(t)$$

Remark: In computing the system Hamiltonian, we must in addition consider the contribution to the system field energy coming from the magnetic field. For the TM case under consideration, this energy is

$$H_{sM} = (\mu/2)\int_{box} < |H_\perp(t,r)|^2 > d^3r$$

where

$$H_\perp = (j\omega\epsilon/h^2)\nabla_\perp E_z \times \hat{z}$$

for a fixed frequency and mode or more precisely, with regard to our modal expansion,

$$H_\perp(t,r) = \epsilon.\sum_{mnp} h_{mn}^{-2} Re(j\omega(mnp)c(mnp)exp(j\omega(mnp)t)).\nabla_\perp u_{mnp}(r)$$

We then find that

$$\int_{box} <|H_{\perp}(t,r)|^2> d^3r =$$

$$(\epsilon^2/2)\sum_{mnp} h_{mn}^{-2}|c(mnp)|^2$$

and hence

$$H_{sM} = (\mu\epsilon^2/4)\sum_{mnp} h_{mn}^{-2}|c(mnp)|^2$$

(a) Fermionic fields as system fields interacting with a photonic bath

(b) Fermionic bath. The creation and annihilation operators satisfying CAR's are

$$J_k(t) = \int_0^t (-1)^{\Lambda_k(s)} dA_k(s),$$

$$J_k(t)^* = \int_0^t (-1)^{\Lambda_k(s)} dA_k(s)^*$$

[8] **Quantum filtering of Yang-Mills gauge fields in interaction with a bath**

The Lagrangian density of the Yang-Mills gauge field is

$$L = (-1/4)Tr(F_{\mu\nu}F^{\mu\nu}) = (-1/4)F_{\mu\nu}^a F^{\mu\nu a}$$

where

$$F_{\mu\nu}^a = A_{\nu,\mu}^a - A_{\mu,\nu}^a + eC(abc)A_\mu^b A_\nu^c$$

We have

$$F_{0r}^a = A_{r,0}^a - A_{0,r}^a + eC(abc)A_r^b A_0^c$$

$$L = (-1/2)F_{0r}^a F_{0r}^a - (1/4)F_{rs}^a F_{rs}^a$$

so the canonical momentum corresponding to the position field A_r^a is

$$P_r^a = \partial L/\partial A_{r,0}^a = -F_{0r}^a$$

Thus,

$$A_{r,0}^a = -P_r^a + A_{0,r}^a - eC(abc)A_r^b A_0^c$$

The Hamiltonian density is then

$$\mathcal{H} = P_r^a A_{r,0}^a - L = -F_{0r}^a A_{r,0}^a + (1/2)F_{0r}^a F_{0r}^a + (1/4)F_{rs}^a F_{rs}^a$$

The Yang-Mills field equations are

$$(D_\mu F^{\mu\nu})^a = 0$$

or more precisely in component form,

$$F_{,\nu}^{\mu\nu a} + C(abc)A_\nu^b F^{\mu\nu c} = 0$$

These field equations can also be expressed in Lie algebra notation as

$$[\nabla_\nu, F^{\mu\nu}] = 0$$

where

$$\nabla_\mu = \partial_\mu - ieA_\mu = \partial_\mu + ieA^a_\mu \tau_a$$

since

$$[\partial_\nu, F^{\mu\nu}] = F^{\mu\nu}_{,\nu},$$

$$[A_\nu, F^{\mu\nu}] = [A^b_\nu \tau_b, F^{\mu\nu c}\tau_c] =$$

$$A^b_\nu F^{\mu\nu c}[\tau_b, \tau_c] = A^b_\nu F^{\mu\nu c} iC(abc)\tau_a$$

[9] **Quantum field theoretic cavity resonator physics using photons, electrons, positrons, non-Abelian gauge Yang mills matter and particle fields and gravitons**

[10] **Quantum control via feedback**

Quantum filtering and control algorithms were first introduced by V.P.Belavkin and perfected by John Gough, Kostler and Lec Bouten.

$$dU(t) = (-(iH + P)dt + L_1 dA - L_2^* dA^* + Sd\Lambda(t))U(t)$$

We take an observable X and note that its Heisenberg evolution is given by

$$j_t(X) = U(t)^* XU(t)$$

Then

$$dj_t(X) = j_t(\theta_0(X))dt + j_t(\theta_1(X))dA(t) + j_t(\theta_2(X))dA(t)^* + j_t(\theta_3(X))d\Lambda(t)$$

where $\theta_k, k = 0, 1, 2$ are linear operators in the linear space of system observables. We take non-demolition measurements in the sense of Belavkin of the form

$$Y_o(t) = U(t)^* Y_i(t)U(t), Y_i(t) = cA(t) + \bar{c}A(t)^* + k.\Lambda(t)$$

The Belavkin filter for this measurement has the following form:

$$\pi_t(X) = \mathbb{E}[j_t(X)|\eta_o(t)], \eta_o(t) = \sigma(Y_o(s) : s \leq t)$$

$$d\pi_t(X) = F_t(X)dt + \sum_{k \geq 1} G_{kt}(X)(dY_o(t))^k$$

where $F_t(X), G_{kt}(X) \in \eta_o(t)$. It is a commutative filter and is therefore called the stochastic Heisenberg equation. Its dual is the stochastic Schrodinger equtation:

$$d\rho_t = F_t^*(\rho_t)dt + \sum_{k \geq 1} G_{kt}^*(\rho_t)(dY_o(t))^k$$

Let $X_d(t)$ be the desired Heisenberg trajectory. Then, the tracking error at time t is $X_d(t) - j_t(X)$. However, we cannot feed this error back into the HP noisy Schrodinger equation because we cannot measure $j_t(X)$ directly without perturbing the system. So we use in place of $j_t(X)$ its real time estimate $\pi_t(X)$ based on the non-demolition measurements $\eta_o(t)$ upto time t and feedback instead the error $X_d(t) - \pi_t(X)$. The system dynamics after feedback is then given by

$$dU(t) = [(-iH + u(t) - P(t))dt + L_1 dA(t) - L_2 dA(t)^* + Sd\Lambda(t))U(t)$$

where

$$u(t) = K(X_d(t) - \pi_t(X))$$

or more generally,

$$u(t) = Lf(X_d(t) - \pi_t(X))$$

We note that

$$dY_o(t) = dY_i(t) + dU(t)^* dY_i(t)U(t) + U(t)^* dY_i(t)dU(t) =$$

$$= dY_i(t) - j(cL_2 + \bar{c}L_2^*)dt + kj_t(S + S^*)d\Lambda + kj_t(L_1 - L_2^*)dA + kj_t(L_1^* - L_2)dA^*$$

So measuring dY_o amounts to measuring $-j_t(cL_2 + \bar{c}L_2^*)dt$ plus noise. In the context of cavity resonator physics, we have that $-L_2$ is the coefficient of dA^* in the HP equation and as we saw, this coefficient is proportional for the kth HP mode to $\sum_n (l_{1kn}(t)c(n) + l_{2kn}(t)c(n)^*)$. This means that our non-demolition measurement corresponds to measuring some projection of the cavity electric field plus noise. Actually, we can construct a whole class of non-demolition measurements that correspond to measuring several projections of the cavity electric field plus noise.

[11] **How to apply machine learning methods to problems in electromagnetics, gravitation and quantum mechanics**

Given an incident em field $(E_i(\omega, r), H_i(\omega, r))$ incident upon a diseased tissue characterized by an inhomogeneous permittivity tensor $\epsilon_{ab}(\omega, r)$ and an inhomogeneous permeability tensor $\mu_{ab}(\omega, r)$, we determine the scattered em fields $(E_s(\omega, r), H_s(\omega, r))$ after it gets scattered by the tissue. The aim is to estimate the permittivity and permeability and derive characteristic features of these using a neural network and match these characteristic features with prototype features to determine the nature of the disease. We train the neural network to take as input a set of incident-scattered field pairs and output the permittivity-permeability parameters. Then when the neural network is presented with another incident-scattered field pair, it will use its trained weights to generate the permittivity and permeability parameters which can be compared with the prototype.

In quantum mechanics, machine learning can be applied as follows: Let $H(\theta)$ be the system Hamiltonian dependent upon an unknown parameter vector to be estimated from repeated measurements of the state taking into account the collapse postulate. The ρ_k denote the state after the kth measurement taken at time t_k and let $\{M_a\}$ denote the POVM. Then, the state after the measurement has been taken at time t_{k+1} is given by

$$\rho_{k+1} = \sum_a \sqrt{M_a} U(t_{k+1} - t_k, \theta) \rho_k U(t_{k+1} - t_k, \theta)^* \sqrt{M_a}$$

if we make the measurement without noting the outcome and if we note the outcome as a, then the state at time t_{k+1} just after the measurement has been made is given by

$$\rho_{k+1}(a) = [\sqrt{M_a} U(t_{k+1} - t_k, \theta) \rho_k U(t_{k+1} - t_k, \theta)^* \sqrt{M_a}]/Tr(numerator)$$

Here

$$U(t, \theta) = exp(-itH(\theta))$$

The joint probability of getting measurement outcomes $a_1, ..., a_k$ respectively at times $t_1 < t_2 < ... < t_k$ is given by

$$Pr(a_1, ..., a_k; t_1, ..., t_k|\theta) =$$

$$Tr(\sqrt{M_{a_k}})$$

[12] Lattice filters and the RLS lattice algorithm:

$$X(n) = [x(n), x(n-1), ..., x(0)]^T,$$

$$X_{n,p} = [z^{-1}X(n), ..., z^{-p}X(n)]$$

$$e_f(n|p) = X(n) - P_{n,p}X(n) = P_{n,p}^{\perp}X(n)$$

where $P_{n,p}$ is the orthogonal projection of $\mathbb{R}^{n+1}$ onto $Range(X_{n,p})$.

$$e_b(n-1|p) = P_{n,p}^{\perp} z^{-p-1}X(n)$$

Let $y(n)$ be another signal. Write

$$Y(n) = [y(n), y(n-1), ..., y(0)]^T$$

Let

$$\hat{Y}(n|p+1) = P_{n+1,p+1}Y(n)$$

Note that $P_{n+1,p+1}$ is the orthogonal projection onto $span\ \{X(n), z^{-1}X(n), ..., z^{-p}X(n)\}$. We can write

$$\hat{Y}(n|p+1) = X_{n+1,p+1}h_{n,p+1}$$

We can also write

$$e_f(n|p) = X(n) + X_{n,p}a_{n,p}, e_b(n-1|p) = z^{-p-1}X(n) + X_{n,p}b_{n-1,p}$$

Update formulas:

$$P_{n,p+1} = P_{n,p} + P_{P_{n,p}^{\perp} z^{-p-1} X(n)}$$

Thus

$$P_{n,p+1}^{\perp} = P_{n,p}^{\perp} - P_{P_{n,p}^{\perp} z^{-p-1} X(n)}$$

$$= P_{n,p}^{\perp} - P_{e_b(n-1|p)}$$

Hence,

$$e_f(n|p+1) = e_f(n|p) - e_b(n-1|p) < e_b(n-1|p), X(n > / \| e_b(n-1|p) \|^2$$

$$= e_f(n|p) - e_b(n-1|p) < e_b(n-1|p), e_f(n|p) > / \| e_b(n-1|p) \|^2$$

$$= e_f(n|p) - K(n|p+1) e_b(n-1|p)$$

from which, it follows that

$$\| e_f(n|p+1) \|^2 = \| e_f(n|p) \|^2 - K(n|p+1)^2 \| e_b(n-1|p) \|^2$$

$$e_b(n|p+1) = e_b(n-1|p) - K(n|p+1) e_f(n|p)$$

and hence

$$\| e_b(n|p+1) \|^2 = \| e_b(n-1|p) \|^2 - K(n|p+1)^2 \| e_f(n|p) \|^2$$

We also easily see using the Gram-Schmidt orthonormalization process that

$$Y(n|p) = X_{n+1,p+1} h_{n,p+1} =$$

$$X(n) h_n(0) + z^{-1} X(n) h_n(1) + ... + z^{-p} h_n(p) =$$

$$e_b(n|0) g_n(0) + e_b(n|1) g_n(1) + ... + e_b(n|p) g_n(p)$$

where

$$g_n(k) = < Y(n), e_b(n|k) >, k = 0, 1, ..., p$$

and hence

$$Y(n|p+1) = Y(n|p) + < Y(n), e_b(n|p+1) > e(b(n|p+1) / \| e_b(n|p+1) \|^2$$

or in other words,

$$g_n(p+1) = < Y(n), e_b(n|p+1) > / \| e_b(n|p+1) \|^2$$

We now look at time updates:

$$X_{n+1,p} = \begin{pmatrix} \xi_{n,p}^T \\ X_{n,p} \end{pmatrix}$$

where

$$\xi_{n,p}^T = [x(n), x(n-1), ..., x(n-p)]$$

Then,

$$a_{n,p} = -(X_{n,p}^T X_{n,p})^{-1} X_{n,p}^T X(n) = -R_{n,p}^{-1} X_{n,p}^T X(n)$$

and

$$R_{n+1,p} = (X_{n+1,p}^T X_{n+1,p}) = R_{n,p} + \xi_{n,p}\xi_{n,p}^T$$

Application of the matrix inversion lemma then gives

RLS lattice algorithm continued
$x[n], n \geq 0$ is a process. Let

$$x_n = [x[n], x[n-1], ..., x[0]]^T \in \mathbb{R}^{n+1},$$

and let

$$z^{-k}x_n = [x[n-k], x[n-k-1], ..., x[0], 0, 0, ..., 0]^T \in \mathbb{R}^{n+1}$$

ie we interpret $z^{-k}x[m] = x[m-k] = 0$, if $k > m$. Forward predictor of order p at time n:

$$e_f(n|p) = P_{n,p}^{\perp} x_n,$$

where $P_{n,p}$ is the orthogonal projection onto $R(X_{n,p})$ where

$$X_{n,p} = [z^{-1}x_n, z^{-2}x_n, ..., z^{-p}x_n] \in \mathbb{R}^{n+1 \times p}$$

Backward predictor of order p at time $n-1$:

$$e_b(n-1|p) = P_{n,p}^{\perp} z^{-p-1} x_n$$

From the basic theory of projection operators in Hilbert space, it follows that

$$e_f(n|p+1) = e_f(n|p) - K_f(n|p) e_b(n-1|p)$$

$$e_b(n|p+1) = \begin{pmatrix} e_b(n-1|p) \\ 0 \end{pmatrix} - K_b(n|p) \begin{pmatrix} e_f(n|p) \\ 0 \end{pmatrix}$$

where

$$K_f(n|p) = < e_f(n|p)k, e_b(n-1|p) > / E_b(n-1|p),$$

$$K_b(n|p) = < e_f(n|p), e_b(n-1|p) > / E_f(n|p),$$

$$E_f(n|p) = \| e_f(n|p) \|^2, E_b(n-1|p) = \| e_b(n-1|p)$$

Then,

$$E_f(n|p+1) = E_f(n|p) - K_f(n|p)^2 E_b(n-1|p)$$

$$E_b(n|p+1) = E_b(n-1|p) - K_b(n|p)^2 E_f(n|p)$$

$$e_f(n|p) = x_n + X_{n,p} a_{n,p}$$

$$e_b(n-1|p) = z^{-p-1} x_n + X_{n,p} b_{n,p}$$

$$a_{n,p} = -R_{n,p}^{-1} X_{n,p}^T x_n, b_{n,p} = -R_{n,p}^{-1} z^{-p-1} x_n$$

where

$$R_{n,p} = X_{n,p}^T X_{n,p}$$

$$X_{n+1,p} = [\xi_{n,p}^T, X_{n,p}^T], \xi_{n,p}^T = [x[n], x[n-1], ..., x[n+1-p]]$$

Also,

$$X_{n,p+1} = [X_{n,p}, z^{-p-1} x_n]$$

Then,

$$R_{n+1,p} = R_{n,p} + \xi_{n,p}\xi_{n,p}^T$$

$$R_{n,p+1} = \begin{pmatrix} R_{n,p} & X_{n,p}^T z^{-p-1} x_n \\ z^{-p-1} x_n^T X_{n,p} & \| z^{-p-1} x_n \|^2 \end{pmatrix}$$

Also,

$$X_{n,p+1} = [z^{-1} x_n, z^{-1} X_{n,p}]$$

and hence

$$X_{n+1,p+1} = [z^{-1} x_{n+1}, z^{-1} X_{n+1,p}]$$

where

$$z^{-1} x_{n+1} = [x[n], x[n-1], ..., x[0], 0]^T = [x_n^T, 0]^T$$

and

$$z^{-1} X_{n+1,p} = [X_{n,p}^T, 0]^T$$

and hence

$$(z^{-1} X_{n+1,p}))^T (z^{-1} X_{n+1,p}) = X_{n,p}^T X_{n,p} = R_{n,p}$$

Then,

$$R_{n+1,p+1} = \begin{pmatrix} \| x_n \|^2 & x_n^T X_{n,p} \\ X_{n,p}^T x_n & R_{n,p} \end{pmatrix}$$

Note: We have that $e_b(n-1|p) = P_{n,p}^{\perp} z^{-p-1} x_n$ and hence $e_b(n|p+1) = P_{n+1,p+1}^{\perp} z^{-p-2} x_{n+1}$ and further,

$$X_{n+1,p+1} = [z^{-1} x_{n+1}, ..., z^{-p-1} x_{n+1}] = [[x_n, z^{-1} x_n, ... z^{-p} x_n]^T, 0]^T$$

$$= [[x_n, X_{n,p}]^T, 0]^T$$

and since $z^{-p-2} x_{n+1} = [z^{-p-1} x_n^T, 0]^T$, we easily see that

$$e_b(n|p+1) = P_{n+1,p+1}^{\perp} z^{-p-2} x_{n+1} = [(P_{R([x_n, X_{n,p}])}^{\perp} z^{-p-1} x_n)^T, 0]^T$$

and

$$P_{R([x_n, X_{n,p}])} = P_{R([e_f(n|p), X_{n,p}])}$$

from which it follows that

$$e_b(n|p+1) = [(P_{n,p}^{\perp} z^{-p-1} x_n - < z^{-p-1} x_n, e_f(n|p) > e_f(n|p)/E_f(n|p))^T, 0]^T$$

$$= [(e_b(n-1|p) - K_b(n|p) e_f(n|p))^T, 0]^T$$

We write

$$A_{n,p}(z) = 1 + \sum_{k=1}^{p} a_{n,p}(k) z^{-k},$$

$$B_{n,p}(z) = z^{-p} + \sum_{k=1}^{p} b_{n,p}(k) z^{-k}$$

Then, we can write

$$e_f(n|p) = A_{n,p}(z)x_n, e_b(n-1|p) = z^{-1}B_{n,p}(z)x_n = B_{n,p}(z)z^{-1}x_n$$

Thus,

$$A_{n,p+1}(z) = A_{n,p}(z) - K_f(n|p)z^{-1}B_{n,p}(z),$$

$$B_{n+1,p+1}(z)x_n = z^{-1}B_{n,p}(z) - K_b(n|p)A_{n,p}(z)$$

Remark:

$$e_b(n|p+1) = z^{-1}B_{n+1,p+1}(z)x_{n+1} = B_{n+1,p+1}(z)[x_n^T, 0]^T$$

$$= [B_{n+1,p+1}(z)x_n^T, 0]^T$$

$$a_{n+1,p+1} = -R_{n+1,p+1}^{-1}X_{n+1,p+1}^T x_{n+1} =$$

$$-\begin{pmatrix} \| x_n \|^2 & x_n^T X_{n,p} \\ X_{n,p}^T x_n & R_{n,p} \end{pmatrix}^{-1} (\xi_{n,p+1}x[n+1] + X_{n,p+1}^T x_n)$$

where

$$\xi_{n,p}^T = [x[n], x[n-1], ..., x[n+1-p]]$$

Note that

$$X_{n+1,p+1} = [\xi_{n,p+1}^T, X_{n,p+1}^T]^T$$

Also,

$$X_{n+1,p+1}^T x_{n+1} = [\xi_{n,p+1}, X_{n,p+1}^T][x[n+1], x_n^T]^T = x[n+1]\xi_{n,p+1} + X_{n,p+1}^T x_n$$

Remark: We now derive a useful formula for the inverse of a block structured symmetric matrix: Let

$$X = \begin{pmatrix} a & \mathbf{b} \\ \mathbf{b}^T & \mathbf{R} \end{pmatrix}$$

where $\mathbf{R}^T = \mathbf{R}$. Writing

$$X^{-1} = \begin{pmatrix} q & \mathbf{p}^T \\ \mathbf{p} & \mathbf{Q} \end{pmatrix}$$

we find that

$$qa + \mathbf{p}^T\mathbf{b} = 1, q\mathbf{b} + \mathbf{p}^T\mathbf{R} = \mathbf{0},$$

$$\mathbf{p}\mathbf{b}^T + \mathbf{Q}\mathbf{R} = \mathbf{I}$$

and hence,

$$q = 1/(a - b^T R^{-1}b))$$

and

$$p = -qR^{-1}b = -(R^{-1}b)/(a - b^T R^{-1}b)$$

$$Q = (I - pb^T)R^{-1} = R^{-1} + R^{-1}bb^T R^{-1}/(a - b^T R^{-1}b)$$

Taking

$$a = \| x_n \|^2, b = X_{n,p}^T x_n, R = R_{n,p}$$

gives us

$$q = (\| x_n \|^2 + x_n^T X_{n,p} a_{n,p})^{-1} = 1/x_n^T(x_n + X_{n,p}a_{n,p}) = 1/E_f(n|p)$$

and

$$p = a_{n,p}/E_f(n|p)$$

$$Q = R_{n,p}^{-1} + R_{n,p}^{-1} X_{n,p} x_n x_n^T X_{n,p} R_{n,p}^{-1}/E_f(n|p)$$

$$= R_{n,p}^{-1} + a_{n,p}a_{n,p}^T/E_f(n|p)$$

Remark:

$$e_f(n|p) = x_n + X_{n,p}a_{n,p}, E_f(n|p) = \| e_f(n|p) \|^2 = x_n^T(x_n + X_{n,p}a_{n,p}) = x_n^T e_f(n|p)$$

Then,

$$a_{n+1,p+1} = [qx[n+1]x_n^T$$

$$R_{n+1,p} = X_{n+1,p}^T X_{n+1,p} =$$

$$\xi_{n,p}\xi_{n,p}^T + R_{n,p}$$

Thus,

$$a_{n+1,p} = -R_{n+1,p}^{-1} X_{n+1,p}^T x_{n+1} =$$

$$= -(R_{n,p}^{-1} - \mu_{n,p}\mu_{n,p}^T/(1 + \eta_{n,p}))(\xi_{n,p}x[n+1] + X_{n,p}^T x_n)$$

$$= a_{n,p} - k_{n,p}(x[n+1] + \xi_{n,p}^T a_{n,p})$$

where

$$\mu_{n,p} = R_{n,p}^{-1}\xi_{n,p}, \xi_{n,p}^T = [x[n], x[n-1], ..., x[n+1-p]],$$

$$\eta_{n,p} = \xi_{n,p}^T R_{n,p}^{-1}\xi_{n,p} = \xi_{n,p}^T \mu_{n,p},$$

$$k_{n,p} = \mu_{n,p}/(1 + \eta_{n,p})$$

$$\mu_{n+1,p+1} = R_{n+1,p+1}^{-1}\xi_{n+1,p+1} =$$

$$\begin{pmatrix} 1/E_f(n|p) & a_{n,p}^T/E_f(n|p) \\ a_{n,p}/E_f(n|p) & R_{n,p}^{-1} + a_{n,p}a_{n,p}^T/E_f(n|p) \end{pmatrix} \begin{pmatrix} x[n+1] \\ \xi_{n,p} \end{pmatrix}$$

$$= \begin{pmatrix} x[n+1]/E_f(n|p) + a_{n,p}^T\xi_{n,p}/E_f(n|p) \\ a_{n,p}x[n+1]/E_f(n|p) + \mu_{n,p} \end{pmatrix}$$

RLS lattice algorithm continued:

$$e_f(n|p+1) = e_f(n|p) - K_f(n|p)e_b(n-1|p), e_b(n|p+1) = [\tilde{e}_b(n|p+1)^T, 0]^T$$

where

$$e_b(n|p+1) = e_b(n-1|p) - K_b(n|p)e_f(n|p)$$

Then,

$$e_f(n|p) = x_n + X_{n,p}a_{n,p}, e_b(n-1|p) = z^{-p-1}x_n + X_{n,p}b_{n-1,p}$$

$$a_{n,p} = -R_{n,p}^{-1}X_{n,p}^T x_n, b_{n-1,p} = -R_{n,p}^{-1}X_{n,p}^T z^{-p-1}x_n,$$

$$R_{n,p} = X_{n,p}^T X_{n,p}$$

$$R_{n+1,p} = \xi_{n,p}\xi_{n,p}^T + R_{n,p}$$

$$R_{n+1,p}^{-1} = R_{n,p}^{-1} - \mu_{n,p}\mu_{n,p}^T/(1+\eta_{n,p})$$

Note that

$$X_{n+1,p} = \begin{pmatrix} \xi_{n,p}^T \\ X_{n,p} \end{pmatrix}$$

$$\mu_{n,p} = R_{n,p}^{-1}\xi_{n,p}$$

$$a_{n+1,p} = -R_{n+1,p}^{-1}X_{n+1,p}^T x_{n+1} =$$

$$-(R_{n,p}^{-1} - \mu_{n,p}\mu_{n,p}^T/(1+\eta_{n,p}))(\mathbf{x}_{n,p}x[n+1] + X_{n,p}^T x_n)$$

$$= a_{n,p} - (\mu_{n,p}/(1+\eta_{n,p}))(x[n+1] + \xi_{n,p}^T a_{n,p})$$

$$= a_{n,p} - k_{n,p}e_f(n+1|n,p)$$

$$\mu_{n+1,p+1} = R_{n+1,p+1}^{-1}\xi_{n+1,p+1}$$

$$R_{n+1,p+1} = \begin{pmatrix} \| x_n \|^2 & x_n^T X_{n,p} \\ X_{n,p}^T x_n & R_{n,p} \end{pmatrix}$$

$$R_{n+1,p+1}^{-1} = \begin{pmatrix} 1/E_f(n|p) & a_{n,p}^T/E_f(n|p) \\ a_{n,p}/E_f(n|p) & (R_{n,p}^{-1} + a_{n,p}a_{n,p}^T)/E_f(n|p) \end{pmatrix}$$

RLS lattice continued:

$$a_{n+1,p} = a_{n,p} - k_{n,p}e_f(n+1|n,p)$$

where

$$e_f(n+1|n,p) = x[n+1] + \xi_{n,p}^T a_{n,p}$$

Note that

$$e_f(n|n,p) = x[n] + \xi_{n-1,p}^T a_{n,p}$$

Is the first component of $e_f(n|p)$. $e_f(n+1|n,p)$ is the first component of $e_f(n+1|p)$ but computed using the filter coefficients $a_{n,p}$ which are estimate on data collected only upto the previous time n.

We have seen that

$$k_{n,p} = \mu_{n,p}/(1+\eta_{n,p}), \mu_{n,p} = R_{n,p}^{-1}\xi_{n,p}/(1+\xi_{n,p}^T R_{n,p}^{-1}\xi_{n,p})$$

We have also derived a time-order update formula relating $\mu_{n+1,p+1}$ with $\mu_{n,p}$.

This formula is given by

$$\mu_{n+1,p+1} = [e_f(n+1|n,p)/E_f(n|p),$$

$$(\mu_{n,p}+E_f(n|p)^{-1}(a_{n,p}a_{n,p}^T\xi_{n,p}-x[n+1]R_{n,p}^{-1}X_{n,p}^T x_n))^T]^T$$

$$= [e_f(n+1|n,p)/E_f(n|p), \mu_{n,p}^T + E_f(n|p)^{-1}e_f(n+1|n,p)a_{n,p}^T]^T$$

The whole logic of the RLS lattice algorithm is to keep computing order and time updates for each of the variables that occur in the process so that finally the recursion closes, ie, gets completed. If at any stage, the recursion does not close, we introduce new variables and then compute order and time updates for the newly introduced variables. This process goes on until finally, the algorithm closes upon itself. We find that

$$E_f(n+1|p) = \| e_f(n+1|p) \|^2 = \| x_{n+1} + X_{n+1,p}a_{n+1,p} \|^2 =$$

$$e_f(n+1|p) = x_{n+1} + X_{n+1,p}(a_{n,p} - k_{n,p}e_f(n+1|n,p))$$

$$= [e_f(n+1|n,p) - \xi_{n,p}^T k_{n,p}e_f(n+1|n,p), (e_f(n|p) - X_{n,p}k_{n,p}e_f(n+1|n,p))^T]^T$$

$$= [(1+\eta_{n,p})^{-1}e_f(n+1|n,p), (e_f(n|p) - X_{n,p}k_{n,p}e_f(n+1|n,p))^T]^T$$

Thus, taking the norm square on both the sides and using the fact that $e_f(n|p)$ is orthogonal to $R(X_{n,p})$ gives us

$$E_f(n+1|p) = e_f(n+1|n,p)^2(1+\eta_{n,p})^{-2} + E_f(n|p) + e_f(n+1|n,p)^2\eta_{n,p}/(1+\eta_{n,p})^2$$

$$= e_f(n+1|n,p)^2/(1+\eta_{n,p}) + E_f(n|p)$$

We now repeat this analysis for the backward prediction errors and filter coefficients. First observe that

$$K_f(n+1|p) = < e_f(n+1|p), e_b(n|p) > /E_f(n+1|p)$$

$$e_b(n|p) = P_{X_{n+1,p}}^{\perp} z^{-p-1}x_{n+1}$$

$$e_f(n|p+1) = x_n + X_{n,p+1}a_{n,p+1} = x_n + [X_{n,p}, z^{-p-1}x_n]a_{n,p+1} =$$

$$e_f(n|p) - K_f(n|p)e_b(n-1|p)$$

$$= (x_n + X_{n,p}a_{n,p}) - K_f(n|p)(z^{-p-1}x_n + X_{n,p}b_{n-1,p})$$

$$= x_n + [X_{n,p}, z^{-p-1}x_n](a_{n,p}^T - K_f(n|p)b_{n-1,p}^T)^T, -K_f(n|p)]^T$$

and hence,

$$a_{n,p+1} = \begin{pmatrix} a_{n,p} - K_f(n|p)b_{n-1,p} \\ -K_f(n|p) \end{pmatrix}$$

Likewise,

$$e_b(n|p+1) = z^{-p-2}x_{n+1} + X_{n+1,p+1}b_{n,p+1} =$$

$$\begin{pmatrix} z^{-p-1}x_n + [x_n, X_{n,p}]b_{n,p+1} \\ 0 \end{pmatrix}$$

$$= \begin{pmatrix} e_b(n-1|p) - K_b(n|p)e_f(n|p) \\ 0 \end{pmatrix}$$

$$= \begin{pmatrix} z^{-p-1}x_n + X_{n,p}b_{n-1,p} - K_b(n|p)(x_n + X_{n,p}a_{n,p}) \\ 0 \end{pmatrix}$$

and hence

$$b_{n,p+1} = \begin{pmatrix} -K_b(n|p) \\ b_{n-1,p} - K_b(n|p)a_{n,p} \end{pmatrix}$$

These represent respectively the order update formulas for the forward and backward prediction filter coefficients. We have computed in addition the time update relation for the forward prediction filter coefficients. We now do this for the backward prediction filter coefficients.

$$b_{n-1,p} = -R_{n,p}^{-1}X_{n,p}^T z^{-p-1}x_n$$

$$b_{n,p} = -R_{n+1,p}^{-1}X_{n+1,p}^T z^{-p-1}x_{n+1} =$$

$$= -(R_{n,p}^{-1} + \mu_{n,p}\mu_{n,p}^T/(1+\eta_{n,p}))(\xi_{n,p}x[n-p] + X_{n,p}^T z^{-p-1}x_n)$$

$$= b_{n-1,p} - k_{n,p}(x[n-p] + \xi_{n,p}^T b_{n-1,p})$$

$$= b_{n-1,p} - k_{n,p}e_b(n|n-1,p)$$

We now need to compute $E_b(n|p)$ in terms of $E_b(n-1|p)$ and also $N(n+1|p) = < e_f(n+1|p), e_b(n|p) >$ in terms of $N(n|p)$. We recall that

$$e_f(n+1|p) = [(1+\eta_{n,p})^{-1}e_f(n+1|n,p), (e_f(n|p) - X_{n,p}k_{n,p}e_f(n+1|n,p))^T]^T$$

and

$$e_b(n|p) = z^{-p-1}x_{n+1} + X_{n+1,p}b_{n,p} =$$

$$\begin{pmatrix} x[n-p] + \xi_{n,p}^T(b_{n-1,p} - k_{n,p}e_b(n|n-1,p)) \\ z^{-p-1}x_n + X_{n,p}(b_{n-1,p} - k_{n,p}e_b(n|n-1,p)) \end{pmatrix}$$

$$= \begin{pmatrix} (1+\eta_{n,p})^{-1}e_b(n|n-1,p) \\ e_b(n-1|p) - X_{n,p}k_{n,p}e_b(n|n-1,p) \end{pmatrix}$$

It follows then by forming the inner product of these two relations that

$$N(n+1|p) = (1+\eta_{n,p})^{-2}e_f(n+1|n,p)e_b(n|n-1,p) + < e_f(n|p)$$

$$-X_{n,p}k_{n,p}e_f(n+1|n,p), e_b(n-1|p) - X_{n,p}k_{n,p}e_b(n|n-1,p) >$$

$$= (1+\eta_{n,p})^{-2}e_f(n+1|n,p)e_b(n|n-1,p) + N(n|p) + (\eta_{n,p}/(1+\eta_{n,p})^2)e_f(n+1|n,p)e_b(n|n-1,p)$$

$$e_f(n+1|n,p)e_b(n|n-1,p)/(1+\eta_{n,p}) + N(n|p)$$

Note that we have used the orthogonality relations

$$X_{n,p}^T e_f(n|p) = X_{n,p}^T e_b(n-1|p) = 0$$

[13] **On a gadget constructed by my colleague Professor Dhananjay Gadre**

In the standard undergraduate course for engineering students called "Signals and Systems", as well as in the associated laboratory courses, the students are taught about how to generate various kinds of periodic signals like the sinewave, the square wave, the ramp wave etc., how to construct the Fourier series of such periodic signals as linear superspositions of higher harmonics of a fundamental frequency and how to design and analyze lowpass, highpass and bandpass filters that would filter such periodic signals so as to retain only a certain small subset of the signal harmonic components. This problem is important, for example, in a situation wherein a circuit designed on a bread-board which receives its input from a power supply gets the current/voltage across its components corrupted by the fundamental as well as higher harmonics of the basic 220-V AC voltage source. This happens because the power supply runs on the AC source and hence some part of the latter signals enter into the circuitry of the power supply. As a result, when we measure the voltage across any of the circuit's elements using a CRO, we observe stray components coming from the AC source and its higher harmonics on the CRO screen. These harmonics are not desirable and hence may be regarded as constituting a component of the noise in addition to the thermal noise being produced in the resistors of the circuit. One way to get rid of these harmonics is to place a filter between the circuit and the CRO to eliminate these stray components. The trouble with such a method is that the filter will consume some of the current and will therefore act as an undesirable load. It would thus be better to have a CRO which automatically does the filtering and hence gives a faithful representation of the circuit's behaviour in the absence of the AC source fundamental and higher harmonic disturbance. Another example where such a filtering is required during the signal recording process is a telephone line in which speakers A and B communicate across a line and due to defects in the line, A hears not only B's speech transmitted over the line but also his own echo. An echo canceller at A's end will use a filter $H(z)$ which may even made adaptive and which takes A's original speech as input and predicts the signal received by A over the line (namely B's speech plus A's own echo). Since A's speech is correlated with only A's own echo and not with B's speech, the filter $H(z)$ will predict only A's echo component in the total signal received by A. Thus, when the filter's output is subtracted from the signal received by A, the result is that A gets the signal spoken by B with a major part of his own echo cancelled out. If we have a CRO that passes through only B's frequency components and rejects A's frequency components (this is possible only when A's and B's speech signals occupy non-overlapping frequency bands. If not, we can shift the band of frequencies occupied by B's speech via appropriate modulation at B's end and use a line which supports both the band of frequencies spoken by A and B's band shifted in frequency via modulation). Then, by just recording the signal received by A at his end using such a CRO, A can get to know B's speech with his own echo removed.

Another application of such a frequency sensitive CRO is in generating sine waves from a square wave. Square waves are easily generated using a switch in a series circuit either turned on and off after fixed durations of time manually or by a rotating motor. A square wave contains all the harmonics of a fundamental and hence a CRO which passes the first N harmonics with variable N can be used to demonstrate Gibb's phenomenon on the behaviour of the sequence of partial sums of a Fourier series near a discontinuity of the signal. Specifically, we can demonstrate that at the discontinuity point of a square wave, the Fourier series converges to around 1.8 times its amplitude. A CRO with variable bandwidth can in fact be used to demonstrate all kinds of behaviour of the partial sums of Fourier series including the fact that at a discontinuity point of the signal, the partial sums converge to the average of the signal amplitude at the immediate left and at the immediate right of the signal.

A CRO at the quantum scale level can also be used to determine the energy levels of an atomic system. Specifically, if $|n>, n = 0, 1, 2, ...$ are the energy eigenstates of an atom with corresponding energy levels $E_n, n = 0, 1, 2, ...$ respectively, then it is known from Schrodinger's equation that if the initial state of the atom is the superposition $|\psi(0)>= \sum_n c(n)|n>$, then after time t, its state will be $|\psi(t)>= \sum_n c(n)exp(-i\omega(n)t)|n>$ where $\omega(n) = 2\pi E_n/h$, h being Planck's constant. Thus, if X is an observable of the atomic system, after time t, its average value in this state will be given by

$$<X>(t) =<\psi(t)|X|\psi(t)>= \sum_{n,m} \bar{c}(n)c(m) <n|X|m> exp(i(\omega(n) - \omega(m))t)$$

when this signal is input into the quantum/nano CRO of the kind described above, it retains only a small finite subset of the frequencies $\omega(n) - \omega(m)$ and by adjusting the bandwidth of the CRO, we can thus determine from spectral analysis of the signal appearing on the CRO, what exactly are the energy level's of the atom or more precisely, what frequencies of radiation can be absorbed or emitted by the atom during transitions caused by perturbing the atom with an external radiation field. Another situation in quantum mechanics that finds application here is related to the notion of repeated measurement and state collapse. Suppose a quantum system has initial state $\rho(0)$. It evolves for time t_1 under the Hamiltonian H to the state $\rho(t_1-) = U(t_1)\rho(0)U(t_1)^*$ where $U(t) = exp(-itH)$. Then a measurement is made using the POVM $M = \{M_a : a = 1, 2, ..., N\}$. If $a(1)$ is the noted outcome, after taking this measurement, the state collapses to

$$\rho(t_1+) = \sqrt{M_{a(1)}}\rho(t_1-)\sqrt{M_{a(1)}})/Tr(\rho(t_1-)M_{a(1)})$$

Again the system evolves for a duration $t_2 - t_1$ to the state

$$\rho(t_2-) = U(t_2 - t_1)\rho(t_1+)U(t_2 - t_1)^*$$

and again the measurement M is made and if the noted outcome is $a(2)$, the state collapses to

$$\rho(t_2+) = \sqrt{M_{a(2)}}\rho(t_2-)\sqrt{M_{a(2)}}/Tr(\rho(t_2-)M_{a(2)})$$

It is then clear that after N such operations, namely, free evolution under the Hamiltonian H followed by applying the measurement M and noting the outcomes at each state, the final state of the atom $\rho(t_N+)$ is expressible as a ratio of two terms. The denominator is a real number and is of the form

$$\sum_{n_1,...,n_N} C(n_1,...,n_N;m_1,...,m_N)exp(-i(\omega(n_1)-\omega(m_1))t_1+(\omega(n_2)$$
$$-\omega(m_2))(t_2-t_1)+...(\omega(n_N)-\omega(m_N))(t_N-t_{N-1})))$$

where $C(n_1,...,n_N)$ are complex numbers while the numerator is of the same form but with the $C(n_1,...,n_N)$ being operators. In other words, the numerator and denominator are multidimensional sinusoids in the variables $(t_1,...,t_N)$ and if we have a generalized spectrum analyzer for multidimensional signals, then we could determine by measuring after time N, the average of an observable in the system state, the frequencies $\omega(n)$, ie, the energy levels of the atomic system Hamiltonian H as well as the initial state $\rho(0)$ and the structure of the measurement operators M_a. In fact, more can be said about this problem. The joint probability of getting $a(1),...,a(N)$ during the above measurement process at times $t_1,...,t_N$ is given simply by

$$P(a(1),...,a(N)|t_1,...,t_N) =$$

$$Tr(E(a(N))U(t_N-t_{N-1})...E(a(2))U(t_2-t_1)E(a(1))U(t_1)\rho(0)U(t_1)^*E(a(1))$$
$$U(t_2-t_1)^*E(a(2)...U(t_N-t_{N-1})E(a(n))$$

where

$$E(a) = \sqrt{M_a}$$

This joint probability is clearly a superposition of multidimensional sinusoids with frequeny-tuples $(\omega(n_1), \omega(n_1) - \omega(n_2), ..., \omega(n_N) - \omega(n_{N-1}))$ with complex amplitudes and a harmonic analyzer of multidimensional sinusoids will be able to determine these frequencies and hence estimate the atomic energy levels.

[14] Summary of the research carried out at the NSUT on design of DSP filters using transmission line elements, design of water antennas, design of antennas based on microstrip cavities of arbitrary cross sectional shape, design of antennas using microstrip cavities filled with material having inhomogeneous permittivity and permeability and design of fractional order filters using transmission line elements.

Acknowledgements: Informal discussions with Dr.Mridul Gupta.

Introduction: A lossless transmission line has the natural property of being able to generate transfer functions with arbitrary fractional delay. The reason for this is that if the input forward and backward voltage amplitudes to the line are V_{1+} and V_{1-} and the corresponding output forward and backward amplitudes are V_{2+}, V_{2-}, then if d is the line length, and $\beta = \omega/u, u = \sqrt{LC}$, then then elementary transmission line analysis shows that

$$V_{2-} = V_{1+}exp(-j\beta d), V_{2+} = V_{1-}exp(j\beta d)$$

or equivalently, if R_0 is the characteristic line impedance, then the voltage and current at the input and output terminals are related by using

$$V_1 = (V_{1+} + V_{1-}), I_1 = (V_{1+} - V_{1-})/R_0,$$

$$V_2 = (V_{2+} + V_{2-}), I_2 = (V_{2+} - V_{2-})/R_0$$

Working in the wave domain, ie, in terms of forward and backward voltage wave components, we have that

$$\begin{pmatrix} V_{2+} \\ V_{2-} \end{pmatrix} = \begin{pmatrix} 0 & exp(j\beta d) \\ exp(-j\beta d) & 0 \end{pmatrix} \begin{pmatrix} V_{1+} \\ V_{1-} \end{pmatrix}$$

since $\beta = \omega/u$, the factor $exp(-j\beta d)$ corresponds to a delay by d/u seconds and hence if time is discretized into steps of Δ seconds, the factor $exp(-j\beta d)$ produces a delay of $d/u\Delta$ samples which need not be an integer if d chosen appropriately. Thus it becomes evident that by connecting in tandem several such Tx line units in conjunction with stub loads, lines having any transfer function with numerator and denominators being superpositions of arbitary fractional powers of the unit delay Z^{-1} can be synthesized. Specifically, consider connecting in parallel to this line of length d, an impedance Z_1 at the input end. Then, the voltage at the input end remains unchanged for fixed voltage and current at the output end while the input current gets modified to $I_1' = I_1 + V_1/Z_1$. We then find that for fixed V_{2+}, V_{2-}, V_{1+}, V_{1-} get modified respectively to

$$V_{1+}' = (V_1 + I_1'R_0)/2 = (V_1 + (I_1 + V_1/Z_1)R_0)/2$$

and

$$V_{1-} = (V_1 - I_1'R_0)/2 = (V_1 - (I_1 + V_1/Z_1)R_0)/2$$

or equivalently

$$V_{1+}' = (V_{1+} + V_{1-})(1 + R_0/Z_1)/2 + (V_{1+} - V_{1-})/2$$

$$= (1 + R_0/2Z_1)V_{1+} + (R_0/2Z_1)V_{1-}$$

and

$$V_{1-}' = (V_{1+} + V_{1-})(1 - R_0/Z_1)/2 - (V_{1+} - V_{1-})/2$$

$$= -R_0V_{1+}/2Z_1 + V_{1-}(1 - R_0/2Z_1)$$

This means that connecting a load in parallel at the input end amounts to mutiplying the S-matrix given above by the matrix

$$\begin{pmatrix} 1+R_0/2Z_1 & R_0/2Z_1 \\ -R_0/2Z_1 & 1-R_0/2Z_1 \end{pmatrix}^{-1}$$

to its right. Since Z_1 can be any function of $j\omega$, this suggests that arbitrary transfer functions with fractional delay elements can be realized. When we have a sequence of 2×2 matrix transfer functions say $T_k(z), k = 1, 2, ..., N$ connected in tandem, the overall transfer matrix is given by

$$S_N(z) = T_1(z)T_2(z)...T_N(z)$$

or equivalently, in recursive form,

$$S_{N+1}(z)(1,1) = S_N(z)(1,1)T_{N+1}(z)(1,1) + S_N(z)(1,2)T_{N+1}(z)(2,1),$$

$$S_{N+1}(z)(1,2) = S_N(z)(1,1)T_{N+1}(z)(1,2) + S_N(z)(1,2)T_{N+1}(z)(2,2),$$
$$S_{N+1}(z)(2,1) = S_N(z)(2,1)T_{N+1}(z)(1,1) + S_N(z)(2,2)T_{N+1}(z)(2,1),$$
$$S_{N+1}(z)(2,2) = S_N(z)(2,1)T_{N+1}(z)(1,2) + S_N(z)(2,2)T_{N+1}(z)(2,2),$$

Sometimes, it may be easy to solve these difference equations. Usually, one defines the scattering parameter $s_{21}(z)$ corresponding to a given transfer matrix as $V_{2-}/V_{1+}|_{V_{1-}=0}$. This this tells us how much amplitude is transferred from the source to the load without reflection. The scattering matrix S is defined via

$$\begin{pmatrix} V_{1-} \\ V_{2-} \end{pmatrix} = S \begin{pmatrix} V_{1+} \\ V_{2+} \end{pmatrix}$$

The elements of S are related to those of the transfer matrix T as follows:

$$V_{1-} = S(1,1)V_{1+} + S(1,2)V_{2+}, V_{2-} = S(2,1)V_{1+} + S(2,2)V_{2+}$$

$$V_{2+} = T(1,1)V_{1+} + T(1,2)V_{1-}, V_{2-} = T(2,1)V_{1+} + T(2,2)V_{1-}$$

[15] An example in quantum mechanics involving two-port parameters: Let $\mathcal{H}_k, k = 1, 2$ be two Hilbert spaces and let $\rho_k, k = 1, 2$ be two density operators defined in the Hilbert spaces $\mathcal{H}_k, k = 1, 2$ respectively. Let H_1, H_2 be two Hamiltonian operators defined in the Hilbert spaces $\mathcal{H}_1, \mathcal{H}_2$ respectively. Let V_1, V_2 be two self-adjoint operators defined in $\mathcal{H}_1, \mathcal{H}_2$ respectively. Let V_{12} be a self-adjoint operator defined in $\mathcal{H}_1 \otimes \mathcal{H}_2$. Let $\mathcal{H} = \mathcal{H}_1 \otimes \mathcal{H}_2$ be the Hilbert space of a quantum system having time varying Hamiltonian

$$H(t) = H_1 \otimes I_2 + I_1 \otimes H_2 + V_{12} + f_1(t)V_1 + f_2(t)V_2$$

$$= H_0 + f_1(t)V_1 + f_2(t)V_2$$

where

$$H_0 = H_1 \otimes I_2 + I_1 \otimes H_2 + V_{12}$$

is the unperturbed Hamiltonian. Note that the perturbing signals $f_1(t), f_2(t)$ are respectively applied to the first and the second component Hilbert spaces respectively. Then, let the initial state of this quantum system be

$$\rho(0) = \rho_1 \otimes \rho_2$$

After time t, the state of the system is

$$\rho(t) = U(t)\rho(0)U(t)^*$$

where

$$U(t) = T\{exp(-i\int_{-\infty}^{\infty} H(t)dt)\}$$

and if we take two observables X_1, X_2 respectively in the component Hilbert spaces $\mathcal{H}_1, \mathcal{H}_2$, then the averages of these observables at time t are respectively given by

$$< X_1 > (t) = Tr_2(\rho(t)(X_1 \otimes I_2))$$

and

$$< X_2 > (t) = Tr_1(\rho(t)(I_1 \otimes X_2))$$

Problem: Compute these averages upto linear orders in $f_1(t), f_2(t)$ and explain how this defines a two port system. Specifically upto linear orders in f_1, f_2, show that we can write

$$< X_1 > (t) = a_1(t) + \int h_{11}(t,\tau)f_1(\tau)d\tau + \int h_{12}(t,\tau)f_2(\tau)d\tau,$$

$$< X_2 > (t) = a_2(t) + \int h_{21}(t,\tau)f_1(\tau)d\tau + \int h_{22}(t,\tau)f_2(\tau)d\tau$$

where the signals $a_1(t), a_2(t)$ are independent of $f_1(.)$ and $f_2(.)$.

[16] **Detecting whether supersymmetry is broken or not using a deep neural network**

For an Abelian gauge-superfield, a gauge invariant action has the form

$$L_g = c_1 f_{\mu\nu}f^{\mu\nu} + c_2\lambda^T\gamma_5\epsilon\gamma^\mu\partial_\mu/\lambda + c_3 D^2$$

This gauge action (which is always Lorentz and gauge invariant) is supersymmetry invariant only for vectors $[c_1, c_2, c_3]$ belonging to a one dimensional subspace of $\mathbb{R}^3$. The problem of determining whether supersymetry is broken on not thus amounts to estimating these constants from measurements. Suppose we allow the gauge particles and their superpartners to interact with a Dirac electron. Then we would have to add the matter action with this gauge action and hence determine the effect of this gauge perturbation upon the matter action and write down the dynamics of the Dirac electron taking into account these interactions.

From the transition probabilities of the Dirac electron under these gauge perturbations, we can estimate the constants c_1, c_2, c_3 and hence determine whether the gauge action respects supersymmetry or not.

In the non-Abelian gauge situation, the complete supersymmetric Lagrangian density for the matter and gauge fields is given by (Reference:S.Weinberg, "The quantum theory of fields vol.III)

$$L =$$

$$-(D_\mu \phi)_n^*(D^\mu \phi)_n - (1/2)\bar{\psi}_n \gamma^\mu (D_\mu \psi)_n + F_n^* F_n$$

$$-Re((f''(\phi))_{nm}\psi_{nL}^T \epsilon \psi_{nL}) + 2Re((f'(\phi))_n F_n) + c_1 Im(\bar{\psi}_{nL}\lambda_A(t_A)_{nm}\phi_m - c_1 Im(\bar{\psi}_{nR}\lambda_A(t_A)_{nm}\phi_m^*$$

$$-\phi_n^*(t_A)_{nm}\phi_m D_A - \xi_A D_A + (1/2)D_A D_A - (1/4)f_{A\mu\nu}f_A^{\mu\nu} - (1/2)\bar{\lambda}_A \gamma^\mu (D_\mu \lambda)_A + c_2 \epsilon(\mu\nu\rho\sigma) f_A^{\mu\nu} f_A^{\rho\sigma}$$

The field equations are obtained by setting the variational derivatives of L w.r.t. all the fields $\psi_{nL}, \psi_{nR}, \psi_{nL}^*, \psi_{nR}^*, \phi_n, \phi_n^*, F_n, F_n^*, V_{A\mu}, \lambda_A, D_A$ to zero. In particular, we obtain a Dirac equation for ψ_n with the operator ∂_μ replaced by the gauge covariant derivative D_μ with the gauge field being $V_{A\mu}$ that define the gauge field tensor $F_{A\mu\nu}$ and with masses dependent upon the scalar field and sources determined by the coupling between the gaugino field λ_A and the scalar field ϕ_m. The dynamical equations for the gaugino field λ_A are again of the $V_{A\mu}$-gauged Dirac form but with zero masses and with sources determined by coupling terms between the scalar field and the Dirac field. Note that the gauge field $V_{A\mu}$ and the scalar matter field ϕ_n are Bosonic while the Dirac field ψ_n and the gaugino field λ_A are Fermionic. The super-Dirac and gaugino equations have the form

$$(i\gamma^\mu D_\mu - m(\phi))\psi_n = \chi_{1n}(\phi, \lambda_A),$$

$$i\gamma^\mu (D_\mu \lambda)_A = \chi_2(\phi, \psi)$$

The super scalar field equation is now of the $V_{A\mu}$-gauged Klein-Gordon form but with a source term determined by a coupling between the Dirac field and the gaugino field. Specifically, it is of the form

$$D^\mu D_\mu \phi_n = \chi_3(\psi, \lambda_A)$$

Deriving these field equations from the above supersymmetric Lagrangian involves first eliminating the auxiliary fields F_n, D_A by noting that their variational equations yield ordinary linear algebraic equations for them in terms of the scalar field. Finally, the field equations for $V_{A\mu}$ or equivalently, form

$$f_{A\mu\nu} = V_{A\nu,\mu} - V_{A\mu,\nu} + C(ABC)V_{B\mu}V_{C\nu}$$

are given by the usual Yang-mills equations but with source current terms now receiving their contribution from all the three fields:the scalar field, the Dirac field and the gaugino field. These three current terms add up without interference and contribute to the dynamics of the Yang-mills gauge field.

[17] **Review of a paper given by Mridul**

This paper studies a certain kind of periodically forced nonlinear oscillator having a delay term in its dynamics as well as a noisy forcing term bilinearly coupled to its position. The delay terms as well as a white Gaussian noise term are parametrized by small parameters. The periodic forcing term is also bilinearly coupled to the oscillator's position process. The complete dynamics is described by a system of two first order differential equations in phase space. The analysis starts first with linearizing the dynamical system and obtaining the characteristic equation for the roots of the linearized system. This characteristic equation is a quadratic plus an exponential function, the exponential term arising due to the delay term. In short, the characteristic equation is a transcendental equation, not soluble in closed form. The authors apply the implicit function theorem to this characteristic equation obtaining thereby a formula for the sensitivity of the roots (the roots are also called the Lyapunov exponents as they give us the rate at which a small perturbation in the initial conditions diverges exponentially) w.r.t the perturbation parameter. After this brief analysis, the authors assume that the Lyapunov exponent is purely imaginary, ie, the linearized differential equation exhibits purely oscillatory behaviour, and derive closed form exact formulas for the oscillation frequency in terms of the perturbation parameter β. After this, they assume that the process can be expressed as a slowly time varying amplitude modulating a sinusoid (just as in amplitude modulation theory) (they have not stated this but it is implicit in equn.(18)) and assume that the delay affects only the sinusoidal carrier part and not the modulation amplitude. I think this portion of the author's analysis requires a more detailed explanation. With this approximation, the authors are able to get rid of the delay term and thereby obtain a standard Ito-stochastic differential equation without delay for the phase, ie, the position and velocity variables (equn.(22) and (26)). In (31), the authors define a stochastic Lyapunov exponent for the Ito sde solution (3). I think that this portion must also be explained more clearly in terms of the ergodic theorem for Brownian motion. Specifically, if $A(t)$ is the stochastic state transition matrix for a linear sde, then for the purpose of defining stochastic Lyapunov exponents, we must ensure conditions under which the limit of $t^{-1}log(\lambda(t))$ exists as $t \to \infty$ where $\lambda(t)$ is an eigenvalue of $A(t)$. The authors then set up the usual Fokker-Planck equation for the pdf of the solution to Ito's sde obtained by removing the delay term using the amplitude modulation technique mentioned in my report above. They derive from this, the stationary/equilibrium solution (36) for the Fokker-Planck equation. I think at this point that some material on bifurcation theory applied to the transcendental characteristic equation can be added to explain how when the perturbation parameter that governs the strength of the delay term is increased gradually from zero, the Lyapunov exponents will change. This analysis can be carried out using standard perturbation theory for the roots of nonlinear equations. The trouble, however, in using any sort of perturbation theory is to prove the convergence of the perturbation series. In (40), the authors write down a "Melnikov function". In order to make the text at this point self-contained, the authors may explain here, how the Melnikov theory can be used to determine the driving frequency in a system of coupled sde's. Specifically, if there are periodic forcing terms in an sde, then the frequency of the forcing term can be extracted approximately in terms of an integral of some function of the stochastic process. I believe that this is the essence of the Melnikov theory. It would be nice to compute the mean and variance of the frequency estimate using this integral. Further, one can in principle derive maximum likelihood estimators of parameters of sde's using stochastic integral representations for the likelihood function. How does Melnikov's method compare with the optimum MLE method ? Some light can be shed on this problem.

Since although the noise is Gaussian, the system is nonlinear, it is follows that the system output will be non-Gaussian. Hence, if we move into higher orders of perturbation theory and not just the linearized theory, apart from the power spectrum, ie, the second order moments, higher order moments and spectra would become important. The authors may shed some light on this without getting into too much computation.

In conclusion, I feel that the paper contains many new and interesting results and I recommend its publication after some light on the issues mentioned above in my report are moderately clarified. The main novelty of the paper appears to be to provide a first step towards describing chaos in a dynamical system in the presence of stochastic noise.

[18] **Application of the RLS lattice algorithm to the problem of estimating the metric from the world lines of particles in a gravtiational field**

Let $g_{\mu\nu}(x|\theta)$ denote the metric of space time which we intend to estimate. Here, θ is the unknown parameter vector which is to be estimated. We assume that

$$g_{\mu\nu}(x|\theta) = h_{\mu\nu}(x) + \sum_{k=1}^{p} \theta[k]\psi_{\mu\nu k}(x)$$

where $\psi_{\mu\nu k}(x)$ are known test functions, $h_{\mu\nu}(x)$ is the unperturbed metric and the unknown parameters θ are small. The geodesic equations are

$$d^2x^r/d\tau^2 + \Gamma^r_{km}(x)dx^k/d\tau)(dx^m/d\tau) + 2\Gamma^r_{k0}(x)(dx^k/d\tau)(dt/d\tau) + \Gamma^r_{00}(x)(dt/d\tau) = 0$$

and

$$d^2t/d\tau^2 + \Gamma^0_{km}(dx^k/d\tau)(dx^m/d\ tau) + 2\Gamma^0_{k0}(dx^k/d\tau)(dt/d\tau) + \Gamma^0_{00}(dt/d\tau)^2 = 0$$

Now, we write down the geodesic equations using perturbation theory upto first order in the parameters θ. It comes out to be of the form

$$d^2x^r/dt^2 = F^r(dx^k/dt, x^k) + \sum_{k=1}^{p} G^{rk}(dx^k/dt, x^k)\theta[k]$$

or equivalently in vector-matrix notation as

$$d^2\mathbf{r}/dt^2 = \mathbf{f}(d\mathbf{r}/dt, \mathbf{r}, t) + \mathbf{G}(d\mathbf{r}/dt, \mathbf{r}, t)\theta$$

We can further discretize this equation and express it as nonlinear difference equation:

$$\mathbf{r}[n+2] = \mathbf{a}(\mathbf{r}[n+1], \mathbf{r}[n], n) + \mathbf{B}(\mathbf{r}[n+1], \mathbf{r}[n], n)\theta$$

This is a second order vector nonlinear time series model and the parameter vector θ can be estimated via the RLS lattice algorithm based on measurements $\mathbf{r}[n], n = 0, 1, ...,$. Note that the order of this time series model is p which is also the number of test function matrices $\psi_{\mu\nu k}(x), k = 1, 2, ..., p$ that we use. We define the following vector and matrix valued functions of time

$$\xi[N] = ((\mathbf{r}[n+2] - \mathbf{a}(\mathbf{r}[n+1], \mathbf{r}[n])))_{n=1}^{N},$$

$$\mathbf{X}[N, p] = [\mathbf{b}_1[N], \mathbf{b}_2[N], ..., \mathbf{b}_p[N]]$$

where

$$\mathbf{b}_k[N] = (\mathbf{B}_k(\mathbf{r}[n+1], \mathbf{r}[n], n)))_{n=1}^{N}$$

Then the above LIP (linear in parameters) model can be expressed in the form

$$\xi[N] = \mathbf{X}[N, p]\theta$$

and we can derive a recursive least squares with time and order updates of the parameter estimates.

[19] A remark about the supersymmetric proof of the Atiyah-Singer-Patodi theorem.

In general relativity, the covariant derivative of a vector field is calculated using the Christoffel symbols as connection components while on the other hand, in non-Abelian quantum field theory, the covariant derivative is calculated in a Lie algebraic manner, ie, as the commutator between the connection covariant derivative and the vector field expressed as as an element of the Lie algebra. This applies also to quantum gravity wherein the non-Abelian connection covariant derivative is $\nabla_\mu = \partial_\mu + \Gamma_\mu$ where Γ_μ, the spinor connection of the gravitational field is expressed in terms of the Clifford algebra generated by the Dirac gamma matrices and a vector field B_μ is also expressed as a spinor field $B_\mu \gamma^\mu = B_\mu e_a^\mu \gamma^a$. The commutator of these two objects then defines the covariant derivative of the vector field. In a normal coordinate system, we wish to prove that near the origin, the latter definition of the covariant derivative as a commutator is the natural one to choose from when one computes the Laplacian as the square of the Dirac operator and this leads immediately to an expression for this Laplacian in terms of the Riemann curvature tensor from which the Atiyah-Singer index theorem can be obtained.

The Dirac operator is

$$D = \gamma^\mu(x)\nabla_\mu$$

Where

$$\gamma^\mu(x) = \gamma(e^\mu(x)) = \gamma^a e_a^\mu(x)$$

Where

$$\{\gamma^a, \gamma^b\} = 2\eta^{ab}$$

Thus,

$$\{\gamma^\mu(x), \gamma^\nu(x)\} = 2g^{\mu\nu}(x)$$

Since

$$g^{\mu\nu}(x) = \eta^{ab} e_a^\mu(x) e_b^\nu(x)$$

The spinor-gravitational connection is

$$\Gamma_\mu(x) = [\gamma^a, \gamma^b] e_a^\nu e_{b\nu:\mu}$$

The Dirac-spinor gravitational covariant derivative is

$$\nabla_\mu = \partial_\mu + \Gamma_\mu$$

We get for a covariant vector B_α,

$$[\nabla_\mu, B_\alpha \gamma^\alpha] = [\nabla_\mu, B_\alpha e_a^\alpha \gamma^a]$$

$$= [\nabla_\mu, B_a \gamma^a] = [\partial_\mu + \Gamma_\mu, B_a \gamma^a]$$

$$= B_{a,\mu} \gamma^a + [\Gamma_\mu, \gamma^a] B_a$$

$$[\Gamma_\mu, \gamma^a] = [\gamma^c \gamma^d \omega_{\mu cd}, \gamma^a]$$

Where

$$\omega_{\mu cd} = e_c^\nu e_{d\nu:\mu}$$

$$= -e_{c:\mu}^\nu e_{d\nu} = -e_{c\nu:\mu} e_d^\nu = -\omega_{\mu dc}$$

$$[\gamma^c \gamma^d, \gamma^a] = 2\eta^{ad} \gamma^c - 2\eta^{ac} \gamma^d$$

Thus,

$$[\Gamma_\mu, \gamma^a] = \omega_{\mu cd} \eta^{ad} \gamma^c From which we deduce that$$

$$[\nabla_\mu, B] = [\nabla_\mu, B_a \gamma^a] =$$

$$B_{a,\mu} \gamma^a + \omega_{\mu cd} \eta^{ad} \gamma^c B_a$$

$$= \gamma^a (B_{a,\mu} + \omega_{\mu ab} \eta^{bc} B_c)$$

Now,

$$2\omega_{\mu ab} = 2e_a^\nu e_{b\nu:\mu} = e_a^\nu e_{b\nu:\mu} - e_b^\nu e_{a\nu:\mu} =$$

$$e_a^\nu (e_{b\nu,\mu} - \Gamma_{\nu\mu}^\alpha e_{b\alpha}) - e_b^\nu (e_{a\nu,\mu} - \Gamma_{\nu\mu}^\alpha e_{a\alpha})$$

Upto $O(x)$, it is clear that

$$e_a^\nu e_{b\nu,\mu} - e_b^\nu e_{a\nu,\mu} = 0$$

Since $e_{a\nu,\mu}$ is $O(x)$ and further, e_a^ν is δ_a^ν upto $O(x)$. Thus, upto $O(x)$, we have that

$$2\omega_{\mu ab} = \Gamma_{\nu\mu}^\alpha (e_{a\alpha} e_b^\nu - e_{b\alpha} e_a^\nu)$$

$$= \Gamma_{ab\mu} - \Gamma_{ba\mu}$$

Since $e_{a\alpha} = \eta_{a\alpha} + O(x)$. Note that $\Gamma_{ab\mu}$ is $O(x)$ and hence

$$\Gamma_{ab\mu}(x) - \Gamma_{ba\mu}(x) = (\Gamma_{ab\mu,m}(0) - \Gamma_{ba\mu,m}(0))x^m + O(x^2)$$

Now,

$$\Gamma_{ab\mu,m} - \Gamma_{ba\mu,m} = (1/2)(g_{ab,\mu m} + g_{a\mu,bm} - g_{b\mu,am} - g_{ba,\mu m} - g_{b\mu,am} + g_{\mu a,bm})$$

$$= (1/2)(g_{a\mu,bm} - g_{b\mu,am} + g_{\mu a,bm} - g_{b\mu,am})$$

Thus,

$$2\omega_{\mu ab}(x) =$$

$$(1/2)(g_{a\mu,bm} - g_{b\mu,am} + g_{\mu a,bm} - g_{b\mu,am})(0)x^m$$

$$= (g_{a\mu,bm}(0) - g_{b\mu,am}(0))x^m$$

This is antisymmetric in (a, b). We wish to show that $(g_{a\mu,bm} - g_{b\mu,am} + g_{\mu a,bm} - g_{b\mu,am})(0)$ for normal coordinates, is also antisymmetric in (μ, m). In fact, for normal coordinates, we have

$$g_{\mu\nu}x^\nu = x^\mu$$

and hence on differentiating,

$$g_{\mu\nu,\rho}x^\nu + g_{\mu\rho} = \delta_{\mu\rho}$$

Another differentiation gives

$$g_{\mu\alpha,\rho} + g_{\mu\nu,\rho\alpha}x^\nu + g_{\mu\rho,\alpha} = 0$$

Another differentiation but now at the point $x = 0$ gives

$$g_{\mu\alpha,\rho\beta}(0) + g_{\mu\beta,\rho\alpha}(0) + g_{\mu\rho,\alpha\beta}(0) = 0$$

Thus

$$g_{a\mu,bm}(0) = -(g_{\mu m,ab}(0) + g_{\mu b,am}(0)$$

and to prove that

$$g_{a\mu,bm}(0) - g_{b\mu,am}(0)$$

is also antisymmetric in (μ, m), we must show that

$$g_{a\mu,bm}(0) - g_{b\mu,am}(0) = -g_{am,b\mu} + g_{bm,a\mu}(0)$$

or equivalently that

$$g_{a\mu,bm} + g_{am,b\mu} - g_{b\mu,am} - g_{bm,a\mu} = 0$$

In view of the above cyclic identity, this amounts to showing that

$$-(g_{ab,m\mu} + g_{am,\mu b}) + g_{am,b\mu} - g_{b\mu,am} - g_{bm,a\mu} = 0$$

or equivalently, that

$$-(g_{ab,m\mu} + g_{b\mu,am} + g_{bm,a\mu}) = 0$$

which is true again by virtue of the cyclic identity. Thus, in normal coordinates, we have that

$$\omega_{\mu ab}(x) = R_{\mu\nu ab}(0)x^{\nu} + O(x^2)$$

This is the fundamental relation that we are looking for.

[20] **Problems in optimal control of quantum fields**

Historical development of quantum filtering and control: Quantum filering and control algorithms in the continuous time case were first introduced by V.P.Belavkin and perfected by John Gough and Lec Bouten. The notion of non-demolition measurements which do not interfere with future state values and which form an Abelian algebra of operators so that one can define in any state, the conditional expection of the state at time t given measurements upto time t were the creation of Belavkin. He based all his computations on the Hudson-Parthasarathy model for the noisy Schrodinger equation and today this approach is the standard recognized way to describe filtering in the quantum context. The main obstacle in quantum filtering was that non-commutativity of observables prevented one from constructing conditional expectations and that was completely and satisfactorily resolved by Belavkin by the introduction of a special class of measurements, namely non-demolition measurements which actually can be shown to be the correct quatnum analogue of the measurement process in the classical filtering theory of Kallianpur and Striebel, namely the measurement diffrential at time t equals the sum of a function of the current system state plus white measurement noise. Lec Bouten in his PhD thesis showed how one can introduce a control term into the Belavkin quantum filter so as to minimize the effect of Lindblad noise in some channels. By Lindblad noise, we mean the noise terms appearing in the unitary evolution of system plus bath in the Hudson-Parthasarathy noisy Schrodinger equation. Later on, other researchers including Belavkin, John Gough and Lec Bouten introduced quantum control in the form of changing the Hamiltonian in the Hudson-Parthasarathy-Schrodinger evolution equation in accordance with a desired output and the actual non-demolition measurement at time t. For example, in the case of a quantum robot, we can alter the Hamiltonian by a torque times angular displacement term where the torque is proportional to the difference between a desired angular momentum and the actual noisy measured angular momentum. Always feedback controllers must be in the form of forces/torques proportional to the difference between a desired output and the actual output or some differential or integral of such a difference or more generally to a linear combination of such terms (like the p.i.d controller in classical control theory). It should be noted that in the quantum context, the parameter in the Hamiltonian being controlled becomes a function of the non-demolition measurement operator.

[1] Let $\rho_s(t)$, the state of a system evolve according to the GKSL equation

$$\rho_s'(t) = -i[H(u(t)), \rho_s(t)] - (1/2)\theta(\rho_s(t), u(t))$$

where $u(t)$ is the coherent state parameter and θ is the Lindblad term with the Lindblad operators L_k dependent upon the coherent state parameter $u(t)$. Specifically, we are assuming that the coherent state $|\phi(u)>$ of the bath slowly varies with time and we write down the Hudson-Parthasarathy qsde for the evolution and then by tracing out over this non-vacuum bath coherent state, obtain the dynamics of the system state:

$$dU(t) = (-(iH + LL^*/2)dt + LdA(t) - L^*dA(t)^*)U(t)$$

$$\rho(t) = U(t)(\rho_s(0) \otimes |\phi(u)><\phi(u)|)U(t)^*,$$

$$\rho_s(t) = Tr_2(\rho(t))$$

We have that

$$d\rho(t) = dU(t)\rho(0)U(t)^* + U(t)\rho(0)dU(t)^* + dU(t)\rho(0)dU(t)^*,$$

$$\rho(0) = \rho_s(0) \otimes |\phi(u)><\phi(u)|$$

Then,

$$Tr_2(dU(t)\rho(0)U(t)^*) =$$

$$Tr_2(Qdt + LdA(t) - L^*dA(t)^*)U(t)\rho(0)U(t)^*)$$

$$= dt.[Q.\rho_s(t) + u(t)L.\rho_s(t) - \bar{u}(t)L^*.\rho_s(t)]$$

where

$$Q = -(iH + LL^*/2)$$

Likewise,

$$Tr_2(U(t)\rho(0)dU(t)^*) = Tr_2(U(t)\rho(0)U(t)^*(Q^*dt + L^*dA(t)^* - LdA(t)))$$

$$= [\rho_s(t)Q^* + \bar{u}(t)\rho_s(t)L^* - u(t)\rho_s(t)L]dt$$

and finally,

$$Tr_2(dU(t)\rho(0)dU(t)^*) = Tr_2[L^*dA(t)^*U(t)\rho(0)U(t)^*LdA(t)]$$

$$= L^*\rho_s(t)Ldt$$

We thus get

$$\rho_s'(t) = [Q.\rho_s(t) + u(t)L.\rho_s(t) - \bar{u}(t)L^*.\rho_s(t)]$$

$$+[\rho_s(t)Q^* + \bar{u}(t)\rho_s(t)L^* - u(t)\rho_s(t)L] + L^*\rho_s(t)L$$

which is precisely the GKSL equation in a non-vacuum coherent state $|\phi(u)>$. We can represent it as

$$\rho_s'(t) = T(\rho_s(t), u(t))$$

where $T(., x)$ is a linear operator on the Banach space of bounded operators in the system Hilbert space parameterized by the complex parameter x. For state tracking, we require to control $u(t)$ so that $\rho_s(t)$ tracks a given state $\rho_d(t)$. This may be termed as coherent quantum control. We can also incorporate conservation processes in to the GKSL dynamics: The relevant Hudson-Parthasarathy noisy Schrodinger equation is

$$dU(t) = (Qdt + L_1 dA(t) - L_2 dA(t)^* + L_3 d\Lambda(t))U(t)$$

where $Q = -(iH + P)$. We then find that

$$Tr_2(dU(t)\rho(0)U(t)^*) =$$

$$[Q\rho_s(t)dt + u(t)L_1\rho_s(t) - \bar{u}(t)L_2\rho_s(t) + |u(t)^2 L_3\rho_s(t)]dt$$

$$Tr_2(\rho(0)dU(t)^*) =$$

$$[\rho_s(t)Q + \bar{u}(t)\rho_s(t)L_1^* - u(t)\rho_s(t)L_2^* + |u(t)|^2\rho_s(t)L_3^*]dt$$

[21] **pde based denoising of quantum image fields**

[1] Let the Lindblad operators L_k be functions of q, p where q, p are canonical position and operator valued n-dimensional vectors. Thus, q is multiplication by the vector x in $\mathbb{R}^n$ and p is the gradient operator $-i\nabla_x$ in $\mathbb{R}^n$. The density operator in position space representation at time t is defined by the Kernel $\rho_t(x,y), x, y \in \mathbb{R}^n$. Thus, for $f \in L^2(\mathbb{R}^n)$, we have that

$$\rho_t f(x) = \int \rho_t(x,y)f(y)dy$$

and

$$\rho_t q f(x) = \int \rho_t(x,y)yf(y)dy$$

$$q\rho_t f(x) = \int x\rho_t(x,y)f(y)dy,$$

$$\rho_t p f(x) = -i\int \rho_t(x,y)\nabla_y f(y)dy = i\int (\nabla_y \rho_t(x,y))f(y)dy$$

$$p\rho_t f(x) = -i\int (\nabla_x \rho_t(x,y))f(y)dy$$

and more generally,

$$q^a p^b \rho_t q^c p^d f(x) = (-i)^{b+d}(-1)^d x^a \int (\nabla_y)^d(\rho_t(x,y)y^c)f(y)dy$$

which is equivalent to saying that the kernel of the operator $q^a p^b \rho_t q^c p^d$ is given by

$$K_t(x,y) = (-i)^{b+d}(-1)^d x^a (\nabla_y)^d(\rho_t(x,y)y^c)$$

In our notation, x^a is an abbreviation for $x_1^{a_1}...x_n^{a_n}$ and p^b is an abbreviation for $p_1^{b_1}...p_n^{b_n}$ or equivalently for

$$(-i)^{b_1+...+b_n}\frac{\partial^{b_1+...+b_n}}{\partial x_1^{b_1}...\partial x_n^{b_n}}$$

More generally, if an operator $L = L(q,p)$ is expressed as a function of q,p with all the $q's$ coming to the left and all the $p's$ to the right, then $L(q,p)\rho_t$ is an opearator having kernel

$$L(x,-i\nabla_x)\rho_t(x,y)$$

while $\rho_t L(q,p)$ has the kernel

$$M(i\nabla_y,y)\rho_t(x,y)$$

where $M(u,v)$ is obtained from $L(u,v)$ by replacing each term of the form $u^r v^s$ in the expansion of $L(u,v)$ by $v^s u^r$. More generally, the kernel of the operator $L_1(q,p)\rho_t L_2(q,p)$ is given by

$$L_1(x,-i\nabla_x)M_2(i\nabla_y,y)\rho_t(x,y)$$

where if

$$L_2(u,v) = \sum_{r,s} a(r,s)u^r v^s$$

then

$$M_2(u,v) = \sum_{r,s} a(r,s)v^s u^r$$

An example of how to derive a Hamiltonian that reproduces the same effect as that of a partial differential operator acting on a quantum field.

Let

$$\psi_k(t,r) = exp(-i(\omega(k)t - k.r))$$

and consider the quantum field

$$X(t,r) = \sum_k (a(k)\psi_k(t,r) + a(k)^*\bar{\psi}_k(t,r))$$

where

$$[a(k),a(j)] = 0, [a(k),a(j)^*] = \delta_{kj}$$

Then

$$i\partial_t\psi_k(t,r) = \omega(k)\psi_k(t,r),$$

$$-i\nabla_r\psi_k(t,r) = k\psi_k(t,r)$$

Hence,

$$\omega(-i\nabla_r)\psi_k(t,r) = \omega(k)\psi_k(t,r)$$

so that ψ_k satisfies the pde

$$i\partial_t\psi_k(t,r) = \omega(-i\nabla_r)\psi_k(t,r)$$

Since we assume $\omega(k)$ to be real valued function, we also get

$$-i\partial_t\bar{\psi}_k(t,r) = \omega(k)\bar{\psi}_k(t,r),$$

$$\omega(i\nabla_r)\bar{\psi}_k(t,r) = \omega(k)\bar{\psi}_k(t,r)$$

and hence

$$-\partial_t^2 X(t,r) = \omega(-i\nabla_r)^2 X(t,r)$$

provided that we assume that

$$\omega(-k) = \omega(k)$$

so that defining the operator 2-vector field

$$Z(t,r) = [X(t,r), i\partial_t X(t,r)]^T$$

we get that

$$i\partial_t Z(t,r) = [i\partial_t X(t,r), \omega(-i\nabla_r)^2 X(t,r)]^T$$
$$= \begin{pmatrix} 0 & 1 \\ \omega(-i\nabla_r)^2 & 0 \end{pmatrix} Z(t,r)$$

This is an example of a quantum field satisfying a vector partial differential equation in space-time with the time derivative being only of the first order.

Let $X(t,r)$ be a quantum image field built out of creation and annihilation operators. We wish To denoise this quantum image field. Let $X_0(t,r)$ denote the corresponding denoised quantum image field. We pass the noisy quantum image field X(t,r) through a spatio-temporal linear filter Having an impulse response $H(t,r)$. The output of this filter is given by

$$\hat{X}_0(t,r) = \int H(t-t', r-r')X(t',r')dt'd^3r'$$

We wish to select the filter $H(t,r)$ so that $\hat{X}_0(t,r)$ is a close approximation to $X_0(t,r)$. in a given quantum state ρ. This means that we select the function $H(t,r)$ so that

$$\int Tr(\rho(X_0(t,r) - \hat{X}_0(t,r))^2)dtd^3r$$

Is minimal. Setting the variational derivative of this error energy function w.r.t. H to zero then gives us the optimal normal equations

$$Tr(\rho \int dtd^3r(X_0(t,r) - \hat{X}_0(t,r))X(t-t', t-r')) = 0$$

or equivalently,

$$\int Tr(\rho\{X_0(t,r), X(t-t', r-r'))\}dtd^3r$$
$$= \int dsd^3uH(s,u) \int Tr(\rho\{X(t-s, r-u), X(t-t', t'-r')\}dtd^3r$$

Thus, to calculate the filter, we must evaluate the symmetrized quantum correlations

$$Tr(\rho\{X_0(t,r), X(t_1,r_1)\}), Tr(\rho\{X(t,r), X(t_1,r_1)\})$$

Assuming that ρ is a quantum Gaussian state so that it is expressible as an exponential of a linear-quadratic form in the creation and annihilation operators $a(k), a(k)^*, k = 1, 2, \ldots$, we express

the quantum fields X_0, X as polynomial functionals in the $a(k), a(k)^*$ and computing the quantum correlations then amounts to calculating the multiple moments of the creation and annihilation operators in a Gaussian state, ie evaluating moments of the form

$$Tr(\rho.\Pi_k(a(k)^{m_k})\Pi_k(a(k)^*n_k))$$

Where

$$\rho = C.exp(-\sum_k \alpha(k)a(k)+\bar{\alpha}(k)a(k)^* - \sum_{k,m} \beta_1(k,m)a(k)a(m)+\beta_2(k,m)a(k)^*a(m)^* \\ -\beta_3(k,m)a(k)^*a(m))$$

The easiest way to evaluate these moments is to use the Glauber-Sudarshan resolution of the identity in terms of coherent states.

[22] Definition of the universal enveloping algebra of a Lie algebra: Let $\mathfrak{g}$ be a Lie algebra and let $(\mathcal{C}, \pi)$ be a pair such that (a) $\mathcal{C}$ is an associative algebra, (b) $\pi : \mathfrak{g} \to \mathcal{C}$ is a linear mapping satisfying $\pi([X,Y]) = \pi(X)\pi(Y) - \pi(Y)\pi(X) \forall X, Y \in \mathfrak{g}$, (c) $\pi(\mathfrak{g})$ generates $\mathcal{C}$ And (d) if $\mathfrak{U}$ is any associative algebra and $\xi : \mathfrak{g} \to \mathfrak{U}$ is a linear map satisfying $\xi([X,Y]) = \xi(X)\xi(Y) - \xi(Y)\xi(X) \forall X, Y \in \mathfrak{g}$, then there exists an algebra homomorphism $\xi' : \mathcal{C} \to \mathfrak{U}$ such that $\xi'(\pi(X)) = \xi(X) \forall X \in \mathfrak{g}$. Then, $(\mathcal{C}, \pi)$ is called a universal enveloping algebra of $\mathfrak{g}$. Theorem: If $(\pi_k, \mathcal{C}_k), k = 1, 2$ are two universal enveloping algebras of a Lie algebra $\mathfrak{g}$, then they are isomorphic in the sense that there exists an algebra isomorphism $\xi : \mathcal{C}_1 \to \mathcal{C}_2$ such that $\xi(\pi_1(X)) = \pi_2(X) \forall X \in \mathfrak{g}$.

[23] **Questions on statistical signal processing**

Attempt any five questions.

[1] Let $X(n), n \in \mathbb{Z}$ be any stationary process with

$$\mathbb{E}(X(n)) = \mu, Cov(X(n), X(m)) = C(n-m)$$

Prove that if

$$C(n) \to 0, |n| \to \infty$$

then

$$lim_{N \to \infty} N^{-1}.\sum_{n=1}^{N} X(n) = \mu$$

in the mean square sense.

[2] If $X(n)$ is a stationary Gaussian process with mean μ and covariance $C(n-m)$, then show that

$$lim_{n \to \infty} N^{-1}.\sum_{n=1}^{N} X(n)X(n+k)$$

converges in the mean square sense to $C(k) + \mu^2$ as $N \to \infty$ provided that $C(n) \to 0$ as $|n| \to \infty$.

[3] Prove the Cesaro theorem: If $a(n)$ is a sequence of complex numbers such that $a(n) \to c$ as $n \to \infty$, then $N^{-1}.\sum_{n=1}^{N} a(n) \to c$ as $N \to \infty$.

[4] Define a filtered probability space and a Martingale and a submartingale in discrete time w.r.t this filtered probability space. Show that if $X(n)$ is submartingale, then the maximal inequality holds:

$$Pr(max_{0 \le n \le N}|X(n)| > \epsilon) \le \mathbb{E}(|X(N)|)/\epsilon$$

Use this result to prove the following version of the strong law of large numbers: If $X(n)$ is a sequence of iid random variables with mean μ and variance σ^2, then almost surely,

$$lim_{N \to \infty} N^{-1}.\sum_{n=1}^{N} X(n) = \mu$$

[5] Write short notes on the following:
(a) The Borel-Cantelli Lemmas.
(b) Application of the Borel-Cantelli lemmas to proving almost sure convergence of a sequence of random variables.
(c) Doob's optional sampling theorem for Martingales and bounded stop-times.
(d) Asymptotic mean and variance of the periodogram of a stationary zero mean Gaussian random process

[6] Define the Ito stochastic integral for an adapted process w.r.t. Brownian motion and prove the existence and uniqueness of solutions to a stochastic differential equation

$$dX(t) = f(X(t))dt + g(X(t))dB(t)$$

interpreted as a stochastic integral equation

$$X(t) - X(0) = \int_0^t f(X(s))ds + \int_0^t g(X(s))dB(s)$$

when f, g satisfy the Lipshitz condition

$$|f(X) - f(Y)| + |g(X) - g(Y)| \le K|X - Y| \forall X, Y \in \mathbb{R}$$

[7] Prove using the Ito calculus that the process

$$M_\lambda(t) = exp(\lambda B(t) - \lambda^2 B(t)/2)$$

is a Brownian martingale. Now apply Doob's optional sampling theorem to this martingale to determine the probability density of the first hitting time of Brownian motion $B(t)$ at a level $a > 0$ given $B(0) = 0$.

[8] Let $x[n]$ be any process. Define the vectors

$$\mathbf{x}_n = [x[n], x[n-1], ..., x[0]]^T \in \mathbb{R}^{n+1}$$

and for $r = 1, 2, ...,$

$$z^{-r}\mathbf{x}_n = [x[n-r], x[n-r-1], ..., x[0], 0, ..., 0]^T \in \mathbb{R}^{n+1}$$

Let

$$\mathbf{a}_{n,p} = [a_{n,p}[1], ..., a_{n,p}[p]]^T$$

denote the optimum forward prediction filter of order p at time n, ie,

$$e_f[n|p] = \mathbf{x}_n + \sum_{k=1}^{p} a_{n,p}[k]z^{-k}\mathbf{x}_n = \mathbf{x}_n + \mathbf{X}_{n,p}\mathbf{a}_{n,p}$$

where

$$\mathbf{X}_{n,p} = [z^{-1}\mathbf{x}_n, ..., z^{-p}\mathbf{x}_n] \in \mathbb{R}^{n+1\times p}$$

is such that

$$E_f(n|p) = \| e_f[n|p] \|^2$$

is a minimum. Likewise, let

$$\mathbf{b}_{n-1,p} = [b_{n-1,p}[1], ..., b_{n-1,p}[p]]^T$$

be the optimum backward prediction filter of order p at time $n-1$, ie, if

$$e_b[n-1|p] = z^{-p-1}\mathbf{x}_n + \sum_{k=1}^{p} b_{n,p}[k]z^{-k}\mathbf{x}_n = z^{-p-1}\mathbf{x}_n + \mathbf{X}_{n,p}\mathbf{b}_{n,p}$$

is such that

$$E_b(n-1|p) = \| e_b[n-1|p] \|^2$$

is a minimum. Derive time and order recursive formulas for the following:

$$e_f(n|p), e_b(n-1|p), \mathbf{a}_{n,p}, \mathbf{b}_{n,p}$$

Also derive time and order projection operator update formulas for the orthgogonal projection

$$\mathbf{P}_{n,p} = \mathbf{X}_{n,p}(\mathbf{X}_{n,p}^T\mathbf{X}_{n,p})^{-1}\mathbf{X}_{n,p}$$

onto $\mathcal{R}(\mathbf{X}_{n,p})$. Now, if $y[n]$ is another process, consider the joint process predictor

$$\hat{\mathbf{y}}_{n,p} = \sum_{k=1}^{p} h_{n,p}[k]z^{-k}\mathbf{x}_n$$

so that

$$\| \mathbf{y}_n - \hat{\mathbf{y}}_{n,p} \|^2$$

is a minimum. Show that

$$\hat{\mathbf{y}}_{n,p} = \mathbf{P}_{n,p}\mathbf{y}_n$$

Show that we can write

$$\mathbf{y}_{n,p} = \sum_{k=0}^{p-1} g_{n,p}[k]e_b[n-1|k]$$

where

$$g_{n,p}[k] = < \mathbf{y}_n, e_b[n-1|k] >, 0 \leq k \leq p-1$$

Derive time an order update formulas for the coefficients $g_{n,p}[k]$. In the process of doing this, you must also derive time and order update formulas for the forward prediction error filter transfer function

$$A_{n,p}(z) = 1 + \sum_{k=1}^{p} a_{n,p}[k]z^{-k},$$

the backward prediction error filter transfer function

$$B_{n,p}(z) = z^{-p-1} + \sum_{k=1}^{p} b_{n,p}[k]z^{-k}$$

and the joint process predictor transfer function

$$H_{n,p}(z) = \sum_{k=1}^{p} h_{n,p}[k]z^{-k}$$

[9] Give all the steps for the construction of the L^2-Ito stochastic integral for an adapted process $f(t)$ w.r.t Brownian motion $B(t)$ over a finite time interval $[0,T]$. Also derive the fundamental properties of this integral, namely

$$\mathbb{E}\int_0^T f(t)dB(t) = 0, \mathbb{E}(\int_0^T f(t)dB(t))^2 = \int_0^T \mathbb{E}(f(t)^2)dt$$

[10] Derive the Levinson-Durbin algorithm for the calculating the forward and backward predictors of a stationary process $x[n]$ with autocorrelation $R[n]$ upto order p. Show that per order iteration, you require only $O(p)$ multiplications as compared to $O(p^2)$ required if you were to calculate the order p predictor directly by solving the relevant optimum normal equations.

[11] Let $X[n]$ be a stationary Gaussian process. Show that the entropy rate $\bar{H}(X)$ of this process defined by

$$\bar{H}(X) = lim_{N \to \infty} N^{-1} H(X[N], X[N-1], ..., X[1])$$

exists and in fact equals

$$H(X(0)|X(-1), X(-2), ...)$$

where if U, V are random vectors, then

$$H(U) = -\int f_U(u) ln(f_U(u)) du, H(U|V) = -\int f_{UV}(u,v) ln(f_{U|V}(u|v) dudv$$

Hence, prove that

$$\bar{H}(X) = (1/2\pi) \int_0^{2\pi} S_X(\omega) d\omega$$

where $S_X(\omega)$ is the power spectral density of the process.

[12] Apply Cramer's large deviation principle to compute the optimal asymptotic false alarm error probability rate as the number of iid samples tends to ∞ as the Kullback-Leibler/information theoretic distance between the two probability distributions.

[24] **Questions on transmission lines and waveguides**

Attempt any five questions.

[1] Calculate the capacitance per unit length between two parallel transmission lines of cylindrical shape having radii a, b with a separation of d between their axes. Use the theory of functions of a complex variable to make this computation.

Super-directivity of a quantum antenna: Suppose that the electron-positron field with wave operator field $\psi(t, r)$. The four current density is then given by

$$J^\mu(x) = -e\psi(x)^* \alpha^\mu \psi(x), \alpha^\mu = \gamma^0 \gamma^\mu$$

And the four vector potential generated by this four current density is then given by

$$A^\mu(x) = \int J^\mu(x') G(x - x') d^4x'$$

Where

$$G(x) = (\mu/4\pi)\delta(x^2) = (\mu/4\pi r)\delta(t - |r|/c)$$

Is the causal Green's function for the wave operator. Now using the above formula for the vector potential, the far field four potential has the form

$$A^\mu(t, r) = (\mu/4\pi r) \int J^\mu(t - r/c + \hat{r}.r'/c, r') d^3r'$$

It follows that as a function of frequency, the far field four potential has the angular amplitude pattern

$$B^{\mu}(\omega,r)=\int J^{\mu}(\omega,r')exp(jk\hat{r}.r')d^3r'$$

To evaluate the directional properties of the corresponding power pattern, we must first choose a state $|\eta>$ for the electron-positron system and compute

$$S^{\mu\nu}(\omega,r)=<\eta|B^{\mu}(\omega,r).B^{\nu}(\omega,r)^*|\eta>=$$

$$\int<\eta|J^{\mu}(\omega,r_1)J^{\nu}(\omega,r_2)^*|\eta>exp(jk\hat{r}.(r_1-r_2))d^3r_1d^3r_2$$

In order to obtain superdirectional properties of the radiated field, we must prepare the state $|\eta>$ so that the above quantum average is large when $\mu=\nu$. First observe that in terms of the creation and annihilation operators of the electron-positron field, the Dirac wave operator field is given by

$$\psi(t,r)=\int[u(P,\sigma)a(P,\sigma)exp(-i(E(P)t-P.r))+$$
$$v(P,\sigma)b(P,\sigma)^*exp(i(E(P)t-P.r))]d^3P$$

Where $E(P)=\sqrt{m^2+P^2}$. We then find that the temporal Fourier transform of The four current density $J^{\mu}(t,r)=-e\psi(t,r)^*\alpha^{\mu}\psi(t,r)$ is given by the convolution

$$J^{\mu}(\omega,r)=(-e/2\pi)\int_{\mathbb{R}}\psi(\omega'-\omega,r)\alpha^{\mu}\psi(\omega',r)d\omega'$$

Where $\psi(\omega,r)$, the temporal Fourier transform of $\psi(t,r)$ is given by

$$\psi(\omega,r)=(2\pi)^{-1}\int[u(P,\sigma)a(P,\sigma)exp(iP.r)\delta(\omega-E(P))+$$
$$v(P,\sigma)b(P,\sigma)^*exp(-iP.r)\delta(\omega+E(P))]d^3P$$

In our CDRA case, we have to modify this formula slightly. The possible frequencies of the Dirac field are not a continuum $E(P),P\in\mathbb{R}^3$ but rather a discrete set $\omega(mnp)=E(P(mnp))$ and at a given oscillation frequency $\omega(mnp)$, the Dirac field contributes an amount

$$\psi_{mnp}(\omega,r)=\chi_1(mnp,\mathbf{r})\delta(\omega-\omega(mnp))a(mnp)$$

If we consider the corresponding negative frequency terms also (ie, radiation from both electrons and positrons), then the result is

$$\psi_{mnp}(\omega,r)=\chi_1(mnp,\mathbf{r})\delta(\omega-\omega(mnp))a(mnp)+\chi_2(mnp,\mathbf{r})\delta(\omega+\omega(mnp))b(mnp)^*$$

The result of performing the above convolution is then

$$J^{\mu}(\omega,r)=\chi_1(mnp,r)$$

Periodogram is an inconsistent estimator of the power spectrum

$$\hat{S}_N(\omega) = N^{-1}|\sum_{n=0}^{N-1} X(n)exp(-j\omega n)|^2$$

$X(n)$ is a stationary zero mean Gaussian process. Then, if $R(n)$ is absolutely summable,

$$\mathbb{E}[\hat{S}_N(\omega)] \to S(\omega), N \to \infty$$

$$\mathbb{E}(\hat{S}_N(\omega_1)\hat{S}_N(\omega_2))$$

$$= N^{-2} \sum_{nmkl=0}^{N-1} \mathbb{E}(X(n)X(m)X(k)X(l))exp(-j\omega_1(n-m)-j\omega_2(k-l))$$

Now,

$$\mathbb{E}(X(n)X(m)X(k)X(l)) = R(n-m)R(k-l)+R(n-k)R(m-l)+R(n-l)R(m-k)$$

The first term based on this decomposition converges to $S(\omega_1)S(\omega_2)$. The second term is

$$N^{-2} \sum_{nmkl=0}^{N-1} R(n-k)R(m-l)exp(-j\omega_1(n-m) - j\omega_2(k-l))$$

$$= N^{-2} \sum R(a)R(b)exp(-j\omega_1(k+a-l-b) - j\omega_2(k-l))$$

$$= N^{-2} \sum R(a)R(b)exp(-j(\omega_1+\omega_2)(k-l)).exp(-j\omega_1(a-b))$$

with the summation range of the indices being

$$0 \le k, l, a+k, b+l \le N-1,$$

or equivalently,

$$|a|, |b| \le N-1, max(0, -a) \le k \le min(N-1, N-1-a),$$
$$max(0, -b) \le l \le min(N-1, N-1-b)$$

It is easy to see that this term can be expressed as

$$[\sum_{a=-(N-1)}^{N-1} R(a)exp(-j(\omega_1+\omega_2)(N-a)/2)exp(-j\omega_1 a).sin(\omega(N-|a|-1)/2)/N.sin(\omega/2)]$$
$$\times[a <--- > b, \omega_1 <---> \omega_2]$$

where $[a < - - -b >, \omega_1 < - - - > \omega_2]$ denotes the same as the previous but with the indicated interchanges. This term evidently converges to zero as $N \to \infty$ for positive ω_1, $omega_2$. The third and last term evaluates to

$$\sum_{n,l=0}^{N-1} R(n-l)exp(-j(\omega_1 n - \omega_2 l)). \sum_{m,k=0}^{N-1} R(m-k).exp(j(\omega_1 m - \omega_2 k))$$

[25] **Further topics in statistical signal processing**

[1] Generalize the Levinson-Durbin algorithm for vector valued AR wide sense stationary processes defined by

$$X(t) = -\sum_{k=1}^{p} A_p(k)X(t-k) + W(t)$$

where $X(t) \in \mathbb{R}^M$ and $A(1), ..., A(p) \in \mathbb{R}^{M\times M}$. Show that the optimal normal equations are assuming that $W(t)$ is iid $N(0, I)$ are given by

$$R[k] + \sum_{m=1}^{p} A_p[m]R[k-m] = E[p]\delta[k], k = 0, 1, ..., p$$

where

$$R[k] = \mathbb{E}(X(t)X(t-k)^T) \in \mathbb{R}^{M\times M}$$

Use the block Toeplitz and block centro-symmetry properties of the autocorrelation matrix

$$((R[k-m]))_{1\le k,m\le p} \in \mathbb{R}^{Mp\times Mp}$$

to obtain order recursive solutions for the matrix coefficients $A_p[k], k = 1, 2, ..., p$. Let

$$B_p[k] = A_p[p+1-k], k = 1, 2, ..., p$$

Then, let

$$J_p = \begin{pmatrix} 0 & 0 & 0 & .. & 0 & I_M \\ 0 & 0 & 0 & ... & I_M & 0 \\ ...I_M & 0 & 0 & ... & 0 & 0 \end{pmatrix} \in \mathbb{R}^{Mp\times Mp}$$

Let

$$S_p = ((R[m-k]))_{1\le k,m\le p]} \in \mathbb{R}^{Mp\times Mp}$$

Then

$$J_p S_p J_p = \tilde{S}_p, J_p^2 = I_{Mp}$$

where

$$\tilde{S}_p = ((R[k-m]))_{1\le k,m\le p}$$

Then, let

$$A_p = [I_M, A_p[1], A_p[2], ..., A_p[p]] \in \mathbb{R}^{M\times M(p+1)}$$

Then, the optimal normal equations can be expressed as

$$A_p S_{p+1} = [E[p], 0, ..., 0]$$

Then,

$$A_p J_{p+1} J_{p+1} S_{p+1} J_{p+1} = [E[p], 0, ..., 0] J_{p+1}$$

or equivalently,

$$B_p \tilde{S}_{p+1} = [0, 0, .., E[p]]$$

where

$$B_p = [A_p[p], A_p[p-1], ..., A_p[1], I_M]$$

Thus,

$$B_p[1:p]\tilde{S}_p + R[p:1] = 0$$

Note that

$$B_p[1:p] = A_p[1:p]J_p$$

On the other hand, we have that

$$A_{p+1}S_{p+2} = [E[p+1], 0, ..., 0]$$

or equivalently, defining

$$A_{p+1}[1:p+1] = [A_{p+1}[1], ..., A_{p+1}[p+1]],$$

we deduce that

$$A_{p+1}[1:p]S_{p+1} + A_{p+1}[p+1]R[-p-1:-1] = [E[p+1], 0, ..., 0]$$

where

$$R[-p-1:-1] = [R[-p-1], R[-p], ..., R[-1]]$$

Also,

$$R[1:p+1] + A_{p+1}[1:p+1]S_{p+1} = 0$$

Note that

$$A_{p+1} = [I, A_{p+1}[1], ..., A_{p+1}[p+1]] = [I, A_{p+1}[1:p+1]]$$

and hence

$$A_{p+1}[1:p]S_p + A_{p+1}[p+1]R[-p:-1] + R[1:p] = 0$$

On the other hand, for an arbitrary $M \times M$ matrix K_{p+1}, we have from the above p^{th} order equations,

$$R[1:p] + A_p[1:p]S_p = 0$$

We define $\tilde{B}_p[1:p]$ to be the solution to

$$R[-1:-p]J_p + \tilde{B}_p[1:p]S_p = 0,$$

or equivalently,

$$R[-p:-1] + \tilde{B}_p[1:p]S_p = 0$$

Likewise, we define $\tilde{A}_p[1:p]$ to be the solution to

$$R[-1:-p] + \tilde{A}_p[1:p]\tilde{S}_p = 0$$

Note that

$$\tilde{B}_p[1:p] = \tilde{A}_p[1:p]J_p$$

Then,

$$(A_p[1:p] - K_{p+1}\tilde{B}_p[1:p])S_p = -R[1:p] + K_{p+1}R[-1:-p]J_p$$

It follows by comparing the above equations that if we take K_{p+1} so that

$$K_{p+1}R[-1:-p]J_p = -A_{p+1}[p+1]R[-p:-1]$$

then we are assured that

$$A_{p+1}[1:p] = A_p[1:p] - K_{p+1}\tilde{B}_p[1:p]$$

However, noting that

$$R[-1:-p]J_p = R[-p:-1]$$

the condition reduces to simply

$$K_{p+1} = -A_{p+1}[p+1]$$

Further, we have defined

$$B_p = A_pJ_{p+1} = [A_p[p], ..., A_p[1], I_M] = [B_p[1:p], I_M]$$

so that

$$B_p\tilde{S}_{p+1} = [0, 0, .., E[p]]$$

and this gives us

$$B_p[1:p]\tilde{S}_p + R[p:1] = 0$$

and hence

$$(B_p[1:p] - \tilde{K}_{p+1}\tilde{A}_p[1:p])\tilde{S}_p = -(R[p:1] - \tilde{K}_{p+1}R[-1:-p])$$

On the other hand, we have that

$$B_{p+1}\tilde{S}_{p+2} = [0, ..., 0, E[p+1]]$$

which implies that

$$B_{p+1}[1:p+1]\tilde{S}_{p+1} + R[p+1:1] = 0$$

or equivalently,

$$B_{p+1}[2:p+1]\tilde{S}_p + B_{p+1}[1]R[-1:-p] = -R[p:1]$$

Thus, if we take

$$\tilde{K}_{p+1} = -B_{p+1}[1] = -A_{p+1}[p+1] = K_{p+1}$$

then we would get

$$B_{p+1}[2:p+1] = B_p[1:p] - K_{p+1}\tilde{A}_p[1:p]$$

We also have

$$R[-1:-p]+\tilde{A}_p[1:p]\tilde{S}_p=0$$

so that

$$R[-1:-p-1]+\tilde{A}_{p+1}[1:p+1]\tilde{S}_{p+1}=0$$

which gives

$$\tilde{A}_{p+1}[1:p]\tilde{S}_p+\tilde{A}_{p+1}[p+1]R[p:1]=-R[-1:-p]$$

and further,

$$(\tilde{A}_p[1:p]-L_{p+1}B_p[1:p])\tilde{S}_p=-R[-1:-p]+L_{p+1}R[p:1]$$

so that if we define

$$L_{p+1}=-A_{p+1}[p+1]=K_{p+1}$$

then we would get

$$\tilde{A}_{p+1}[1:p]=\tilde{A}_p[1:p]-K_{p+1}B_p[1:p]$$

[26] Reduce Dirac's relativistic wave equation in a radial potential to the the problem of solving two coupled ordinary second order linear differential equations in a single radial variable r. Solve these by the power series method. Now add quantum noise in the form of superposition of the derivatives of the creation and annihilation processes for the vector potential and superposition of the creation and annihilation processes for the scalar potential (start with the noisy vector potential expressed in terms of creation and annihilation process derivatives and apply the Lorentz gauge condition to derive the expression for the noisy scalar potential). Then incorporate quantum Ito's correction terms in the Hudson-Parthasarathy noisy Dirac equation and assuming the bath to be in a coherent state, obtain formulas upto $O(e^2)$ for the transition probability of the bound Dirac electron subject to quantum noise from one stationary state to another.

[27] Notion of group algebra for a finite group and a representation of the group algebra in terms of a representation of the finite group.

[28] Write down the Lie algebra representations of the standard generators of $SO(3)$ acting on differentiable functions defined on $\mathbb{R}^3$ and show that these generartors are precisely the angular momentum operators in quantum mechanics, ie, the x, y, z components of the operators $-i\mathbf{r}\times\nabla$.

[29] The standard generators of the Lie algebra $\mathfrak{sl}(2,\mathbb{C})$ are denoted by X, Y, H. They satisfy the commutation relations

$$[H,X]=2X,[H,Y]=-2Y,[X,Y]=H$$

From these commutation relations, derive all the finite dimensional irreducible representations of $\mathfrak{sl}(2,\mathbb{C})$ and hence of $SL(2,\mathbb{C})$ and $SU(2,\mathbb{C})$. Show that $SL(2,\mathbb{C})$ is the complexification of $SU(2,\mathbb{C})$ and also of $SL(2,\mathbb{R})$. These are upto isomorphism, the only two non-conjugate real forms of $SL(2,\mathbb{C})$. The former is a compact real form while the latter is a non-compact real form.

[30] Some problems in group representation theory.

[1] Define the Weyl group for a semisimple Lie algebra. First define and prove the existence of a Cartan subalgebra, ie, a maximal Abelian algebra and show that all its elements in the adjoint representation are semi-simple. This is true only for semisimple Lie algebras. The Weyl group is defined as the normailizer of the Cartan subalgebra. Show that the Weyl group is generated by the set of reflections corresponding to simple roots w.r.t. any positive system of roots. Show that the Weyl group acts dually on the set of roots by permuting them. Give examples of complex simple Lie algebras, their Cartan subalgebras and the associated Weyl groups. Consider $sl(n,\mathbb{C}), so(n,\mathbb{C}), sp(n,\mathbb{C})$. Use a convenient basis for these simple Lie algebras for constructing the root vectors and the Cartan subalgebras. Show that every complex semisimple Lie algebra has upto to conjugacy equivalence just one Cartan subalgebra. Show by taking the example of $sl(2,\mathbb{R})$, that real semisimple Lie algebras can have more than one non-conjugate Cartan subalgebra.

[2] Show that given a representation π of a semisimple Lie algebra $\mathfrak{g}$, and a vector v in the vector space V in which the representation π acts, suppose v is a cyclic vector and that $\pi(H)v = \lambda(H)v\forall H \in \mathfrak{h}$, ie, $v \in V_\lambda$ for some $\lambda \in \mathfrak{h}^*$ And $\pi(X_i)v = 0, i = 1,2,\ldots,l$, then if $\mathfrak{N}_-$ is the subalgebra of the universal enveloping algebra $\mathfrak{G}$ of $\mathfrak{g}$. then $V = \pi(\mathfrak{N}_-)v$. Hence deduce that π is a representation with weights, ie, we can write

$$V = \bigoplus_{\mu\in\mathfrak{h}} V_\mu$$

Where

$$V_\mu = \{w \in V : \pi(H)w = \mu(H)w\forall H \in \mathfrak{h}\}$$

Show that

$$dim V_\lambda = 1$$

ie

$$V = \mathbb{C}.v$$

Show further that if π is an irreducible representation, and if $w \in V$ is any vector such that $\pi(X_k)w = 0, 1 \leq k \leq l$, then $w \in V_\lambda$.

[3] Let π be a representation of a Lie algebra $\mathfrak{g}$ in V and let v be a non-zero vector in V. Let $\mathfrak{G}$ denote the universal enveloping algebra of $\mathfrak{g}$ and define

$$\mathfrak{M} = \{a \in \mathfrak{G} : \pi(a)v = 0\}$$

Show that $\mathfrak{M}$ is an ideal in $\mathfrak{G}$. Define a map

$$T : \mathfrak{G}/mathfrakM \to V$$

By

$$T(a + \mathfrak{M}) = \pi(a)v, a \in \mathfrak{G}$$

Show that T is a well defined linear map between vector spaces and that T is one-one. Let $\mathfrak{M}_1$ be an ideal in $\mathfrak{G}$ containing $\mathfrak{M}$. Then, show that $W = T(\mathfrak{M}_1/\mathfrak{M})$ consists of all elements $\pi(a)v, a \in \mathfrak{M}_1$ in V and is therefore a π-invariant subspace of V. Hence deduce that if π is an irreducible representation, then T is a bijection and $\mathfrak{M}_1$ is either $\mathfrak{M}$ or $\mathfrak{G}$, ie, in other words, $\mathfrak{M}$ is a maximal ideal in $\mathfrak{G}$. Conversely, show that If π is an arbitrary representation and $\mathfrak{M}$ as defined above is a maximal ideal of $\mathfrak{G}$ with v as a π-cyclic vector then π is an irreducible representation. In fact, show that if v is π-cyclic for any representation π, then T is a bijection and the mapping

$$\mathfrak{M}_1/\mathfrak{M} \to T(\mathfrak{M}_1)$$

Is a bijection between the set of all maximal ideals $\mathfrak{M}_1$ of $\mathfrak{G}$ containing $\mathfrak{M}$ and the set of all π-invariant subspaces of V.

Give examples of infinite dimensional Lie algebras having finite dimensional non-trivial representations. Show that if G is a simply connected Lie group and H is a discrete normal subgroup, then the Group $\pi_1(G/H)$ is isomorphic to H. Hint: First show that if $\gamma : [0,1] \to G/H$ is any continuous curve and if $p : G \to G/H$ is the canonical projection, then there exists a unique continuous curve $\tilde{\gamma} : [0,1] \to G$ such that $po\tilde{\gamma} = \gamma$. For this, you must make use of the discreteness of H. Now let $\gamma : [0,1] \to G/H$ be a closed curve starting at H. Then, define $M(\gamma) = \tilde{\gamma}(1)$. Prove that if $\mathcal{C}$ is the equivalence class of all Closed curves in G/H starting at H and with composition defined in the usual way that one defines in homotopy, ie $\gamma_1 o \gamma_2(t) = \gamma_1(2t), 0 \le t \le 1/2$ and $= \gamma_2(2t-1)$ for $1/2 \le t \le 1$, then $\mathcal{C}$ becomes a group and that M is an isomorphism from $\mathcal{C}$ to H. To show that, you must make use of the simple connectedness of G. Indeed, let $\gamma \in \mathcal{C}$ and let $\tilde{\gamma} : [0,1] \to G$ be the uniquely defined continuous curve as above. Then, if $\tilde{\gamma}(1) = e$, it follows from the simple connectedness of G that $\tilde{\gamma}$ is homotopic to the constant curve $\tilde{\gamma}_0(t) = e, 0 \le t \le 1$ from which it follows that $\gamma = po\tilde{\gamma}$ is homotopic to $\gamma_0 = po\tilde{\gamma}_0$. This proves the injectivity of M and the surjectivity of M is obvious. Note the way in which we make use of the normality of H in G: Only because H is normal, it follows that G/H is a group owing to which, if we define the composition of two curves $\tilde{\gamma}_1$ and $\tilde{\gamma}_2$ in G both starting at e by the rule $\tilde{\gamma} = \tilde{\gamma}_1 o \tilde{\gamma}_2$, where $\tilde{\gamma}(t) = \tilde{\gamma}_1(t)$ For $0 \le t \le 1/2$ and $\tilde{\gamma}(t) = \tilde{\gamma}(1) o \tilde{\gamma}_2(2t-1), 1/2 \le t \le 1$, then it follows that

$$p(\tilde{\gamma}_1 o \tilde{\gamma}_2) = p(\tilde{\gamma}_1) o p(\tilde{\gamma}_2)$$

[31] **New topics in plasmonic antennas**

[1] Quantum Boltzmann equation, Version 1: We have a creation annihilation operator field in three momentum space $a(K), a(K)^*, K \in \mathbb{R}^3$. These satisfy the canonical commutation relations (CCR):

$$[a(K), a(K')^*] = \delta^3(K - K')$$

Let ρ denote the density matrix of the system (at time $t = 0$). Under the Heisenberg matrix mechanics, ρ remains a constant while $a(K), a(K)^*$ evolve with time to $a(K,t), a(K,t)^*$. Let H denote the Hamiltonian of the system. Then,

$$a(K,t) = exp(itH)a(K).exp(-itH)$$

$$a(K,t)^* = exp(itH)a(K)^*.exp(-itH)$$

Alternately, if we adopt the Schrodinger wave mechanics picture, the operators $a(K), a(K)^*$ remain constant with time while the density ρ evolves with time to

$$\rho(t) = exp(-itH)\rho.exp(itH)$$

We assume that $\rho = exp(-\beta H)/Z(\beta), Z(\beta) = Tr(exp(-\beta H))$, ie, ρ is the canonical Gibbs density. Then, ρ remains constant with time under the Schrodinger picture. Define the Wigner-distribution function

$$f(r,K,t) = Tr(\rho(t).a(K + r/\tau)a(K - r/\tau)^*)$$

We assume that

$$H = \int g(K)a(K)^*a(K)d^3K + \int h(K_1,K_2)a(K_1)^*a(K_1)a(K_2)^*a(K_2)d^3K_1d^3K_2$$

and we take $\rho(0)$ as the Gaussian state

$$\rho(0) = exp(-\beta \int g(K)a(K)^*a(K)d^3K)/Z(\beta)$$

We calculate

$$\partial f(r,K,t)/\partial t = Tr(-i[H,\rho(t)]a(K + r/\tau)a(K - r/\tau)^*)$$

$$= iTr(\rho(t)[H, a(K + r/\tau)a(K - r/\tau)^*])$$

Now,

$$[H, a(K+r/\tau)a(K-r/\tau)^*] = [H, a(K+r/\tau)]a(K-r/\tau)^* + a(K+r/\tau)[H, a(K-r/\tau)^*]$$

Now,

$$[H, a(K + r/\tau)] = \int g(K')[a(K')^*a(K'), a(K + r/\tau)]d^3K'$$

$$+ \int h(K_1,K_2)[a(K_1)^*a(K_1)a(K_2)^*a(K_2), a(K + r/\tau)]d^3K_1d^3K_2$$

Now,

$$[a(K')^*a(K'), a(K+r/\tau)] = [a(K')^*, a(K+r/\tau)]a(K') = -\delta^3(K-K'+r/\tau)a(K'),$$

$$[a(K_1)^*a(K_1)a(K_2)^*a(K_2), a(K+r/\tau)] =$$

$$[a(K_1)^*, a(K+r/\tau)]a(K_1)a(K_2)^*a(K_2)+a(K_1)^*a(K_1)[a(K_2)^*, a(K+r/\tau)]a(K_2)$$

$$= -\delta^3(K-K_1+r/\tau)a(K_1)a(K_2)^*a(K_2) - \delta^3(K-K_2+r/\tau)a(K_1)^*a(K_1)a(K_2)$$

Thus,

$$[H, a(K+r/\tau)] = -\int g(K')\delta^3(K-K'+r/\tau)a(K')d^3K'$$

$$-\int h(K_1, K_2)(\delta^3(K-K_1+r/\tau)a(K_1)a(K_2)^*a(K_2)$$

$$-\delta^3(K-K_2+r/\tau)a(K_1)^*a(K_1)a(K_2))d^3K_1d^3K_2$$

$$= -g(K+r/\tau)a(K+r/\tau) - a(K+r/\tau)\int h(K+r/\tau, K_2))a(K_2)^*a(K_2)d^3K_2$$

$$-(\int h(K_1, K+r/\tau)a(K_1)^*a(K_1)d^3K_1)a(K+r/\tau)$$

$$-g(K+r/\tau)a(K+r/\tau) - \{a(K+r/\tau), \int h(K+r/\tau, K')a(K')^*a(K')d^3K'\}$$

Note that without any loss of generality, we are assuming that

$$h(K_1, K_2) = h(K_2, K_1)$$

[32] **Hartree-Fock equations for an N-electron atom**
The Hamiltonian is given by

$$H = \sum_{k=1}^{N} H_{0k} + \sum_{1\le k<j\le N} V_{kj}$$

where

$$H_{0k} = P_k^2/2m + V_{0k}$$

with

$$V_{0k} = V_0(r_k) + (ge/2m)(\sigma_k, B_k)$$

where

$$V_0(r_k) = -Ze^2/4\pi\epsilon r_k$$

and B_k is the magnetic field produced by the nucleus at the site of the kth electron owing to the its relative motion. It is given by

$$B_k = -\mu e(v_k \times r_k)/4\pi r_k^3 = -e(P_k \times r_k)/4\pi m r_k^3$$

$$= eL_k/4\pi m r_k^3$$

where

$$L_k = r_k \times P_k = -ir_k \times \nabla_k$$

is the angular momentum of the kth electron. Thus,

$$(ge/2m)(\sigma_k, B_k) = ge^2(\sigma_k, L_k)/8\pi m^2 r_k^3$$

This is also called the spin-orbit interaction. Further, the interaction between the kth and the jth electron for $k \neq j$ is given by

$$V_{kj} = e^2/4\pi\epsilon r_{kj} + ge(\sigma_k, B_{kj})/2m$$

where B_{kj} is the magnetic field at the site of the kth electron produced by the jth electron. This magnetic field has two components. One, the magnetic field produced by the spin of the jth electron and two, the magnetic field produced by the relative motion between the two electrons. This first component is given by

$$B_{kj}^{(1)} = curl(\mu m_j \times r_{kj}/4\pi r_{kj}^3), m_j = ge\sigma_j/2m$$

and the second component is given by

$$B_{kj}^{(2)} = -\mu e(P_k - P_j) \times r_{kj}/4\pi m r_{kj}^3$$

An alternate way to describe this interaction is to consider the spin-orbital magnetic moment of the jth electron:

$$M_j = e(L_j + g\sigma_j)/2m$$

and to consider the magnetic field

$$curl((\mu/4\pi)M_j \times r_{kj}/r_{kj}^3)$$

produced by this total magnetic moment at the site of the kth electron. Yet another model for this interaction is to consider the total electromagnetic field produced by the jth electron:

$$E_j(r_k) = -er_{kj}/4\pi\epsilon r_{kj}^3,$$

$$B_j(r_k) = curl((\mu/4\pi)M_j \times r_{kj}/r_{kj}^3)$$

at the site of the kth electron and then to determine this electro-magnetic field at the rest frame of the kth electron:

$$\tilde{E}_j(r_k) = E_j(r_k) + (P_k/m) \times B_j(r_k),$$

$$\tilde{B}_j(r_k) = B_j(r_k)$$

provided that we neglect relativistic effects. Taking all these effects into account, we can express the total Hamiltonian of the system of N electrons in the form

$$H = \sum_k (P_k^2/2m - Ze^2/4\pi\epsilon r_k) + \sum_k f_1(r_k)(\sigma_k, L_k)$$

$$+\sum_{j<k} f_2(r_{kj}(\sigma_k, \sigma_j) + \sum_{j<k} f_3(r_{kj})(L_k, L_j)$$

$$+\sum_{j<k} f_4(r_{kj})(L_k, \sigma_j)$$

To simplify our calculations, we shall specialize these results to a two electron atom, as for example the Helium atom. The total Hamiltonian of this system of two electrons is given by

$$H = H_{01} + H_{02} + V_{12}$$

where

$$H_{01} = P_1^2/2m - Ze^2/r_1 + f_1(r_1)(\sigma_1, L_1), H_{02} = P_2^2/2m - Ze^2/r_2 + f_1(r_2)(\sigma_2, L_2)$$

$$V_{12} = e^2/4\pi\epsilon r_{12} + (e/2m)(B_1(r_2), L_2 + g\sigma_2)$$

where

$$B_1(r_2) = ((\mu e/4\pi m r_{12}^3)(P_2 - P_1) \times r_{21})$$
$$+curl_2((\mu/4\pi r_{12}^3)(ge\sigma_1/2m) \times r_{21})$$

Another way to calculate the interaction between the spin and orbital magnetic moments of the two electrons is to calculate the total magnetic field produced by both the electrons and to integrate its square over the whole of space. The resulting energy is then equal to $(1/2\mu) \int |B_1(r) + B_2(r)|^2 d^3r$ and the interaction component of this energy equals $(1/\mu) \int (B_1(r), B_2(r)) d^3r$ which is proportional to a term of the form $f(r_{12})(L_1 + g\sigma_1, L_2 + g\sigma_2)$ because the magnetic field produced by the first electron is

$$B_1(r) = curl(\mu/4\pi|r - r_1|^3)M_1 \times (r - r_1)), M_1 = e(L_1 + g\sigma_1)/2m$$

and likewise for $B_2(r)$. Thus,

$$V_{12} = e^2/4\pi\epsilon r_{12} + f_2(|r_{12}|)(L_1 + g\sigma_1, L_2 + g\sigma_2)$$

which is of the above form. Thus, we write

$$H = (P_1^2/2m - Ze^2/r_1 + f_1(r_1)(\sigma_1, L_1)) + (P_2^2/r_2 - Ze^2/r_2 + f_1(r_1)(\sigma_2, L_2))$$
$$+(e^2/4\pi\epsilon r_{12}) + f_2(|r_{12}|)(L_1 + g\sigma_1, L_2 + g\sigma_2)$$

In order to apply the variational principle for computing the approximate wave functions of the two electrons, we first note that the test wave function must be antisymmetric with respect to interchange of the position-momentum-spin indices in accordance with the Pauli exclusion principle. Thus we may try wave functions of the form

$$\psi(12) = C_1(\phi_1(r_1)\phi_2(r_2) - \phi_1(r_2)\phi_2(r_1))|++>$$

$$\psi(12) = C_2(\phi_1(r_1)\phi_2(r_2) - \phi_1(r_2)\phi_2(r_1))|-->$$

$$C_3(\phi_1(r_1)\phi_2(r_2)+\phi_1(r_2)\phi_2(r_1))(|+->-|-+>)$$

where ϕ_1 and ϕ_2 are normalized position space wave functions and C_1, C_2, C_3 are normalization constants given by

$$1/C_1^2 = 2(1- <\phi_1,\phi_2>) = 1/C_2^2,$$

$$1/C_3^2 = 4(1+ <\phi_1,\phi_2>)$$

assuming the wave functions to be real.

[33] **Quantum Boltzmann equation for indistinguishable particles in a system in the presence of an external quantum electromagnetic field**

$\rho(t)$ is the state of the system and bath. The total Hamiltonian of the system and bath has the form

$$H = H_s + H_B + V_{sB}$$

where H_s is the system Hamiltonian, H_B the bath field Hamiltonian and V_{SB} is the interaction Hamiltonian between the system and bath. The system Hilbert space is

$$\mathcal{H}_S = \bigotimes_{n=1}^{N} \mathcal{H}_n$$

where $\mathcal{H}_n$ is the Hilbert space of the nth particle in the system and these Hilbert spaces are identical copies. The bath Hilbert space is $\mathcal{H}_B$. H_s acts in the system Hilbert space and is therefore to be identified as $H_s \otimes I_B$. H_B acts in the bath Hilbert space and is therefore to be identified with $I_s \otimes H_B$ while V_{sB} acts in $\mathcal{H}_s \otimes \mathcal{H}_B$ and can therefore be expressed as

$$V_{sB} = \sum_k V_{sk} \otimes V_{Bk}$$

where V_{sk} acts in $\mathcal{H}_s$ while V_{Bk} acts in $\mathcal{H}_B$. The system Hamiltonian has the form

$$H_s = \sum_{n=1}^{N} H_{sn} + \sum_{1\leq n<k\leq N} V_{nk}$$

where H_{sn} acts in $\mathcal{H}_n$, the nth particle Hilbert space and V_{nk} acts in $\mathcal{H}_n \otimes \mathcal{H}_k$. We assume that $H_{sn}, n = 1, 2, ..., N$ are identical copies and so are $V_{nk}, 1 \leq n < k \leq N$. The system state at time t is given by

$$\rho_s(t) = Tr_B(\rho(t))$$

while the bath state at time t is given by

$$\rho_B(t) = Tr_s(\rho(t))$$

The Schrodinger-Liouville-Von-Neumann equation for the density operator is

$$i.rho'(t) = [H, \rho(t)] = [H_s, \rho(t)] + [H_B, \rho(t)] + [V_{sB}, \rho(t)]$$

and taking the partial trace, we get

$$i\rho_s'(t) == [H_s, \rho_s(t)] + Tr_B[V_{sB}, \rho(t)],$$

$$i\rho_B'(t) = [H_B, \rho_B(t)] + Tr_s[V_{sB}, \rho(t)]$$

In order to isolate the system dynamics from the bath dynamice we make an approximation

$$\rho(t) \approx \rho_s(t) \otimes \rho_B(t)$$

while calculating the term $Tr_B[V_{sB}, \rho(t)]$ (Since V_{sB} is already assumed to be of the first order of smallness, this assumption amounts to neglecting second order of smallness terms). Thus, we get for the system state dynamics the approximate equation

$$i\rho_s'(t) = [H_s, \rho_s(t)] + \sum_k Tr(\rho_B(t)V_{Bk})[V_{sk}, \rho_s(t)]$$

Writing

$$Tr(\rho_B(t)V_{Bk}) = a_k(t), V_s(t) = \sum_k a_k(t)V_{sk},$$

we can write

$$i\rho_s'(t) = [H_s, \rho_s(t)] + [V_s(t), \rho_s(t)]$$

where all operators here are now system space operators. We then assume that the particles within the system are all indistinguishble and hence the partial traces of $\rho(t)$ over all the $\mathcal{H}_k's$ except $\mathcal{H}_n$ are identical copies for the different $n's$. We then get on taking partial traces:

$$i\rho_{s1}'(t) = [H_{s1}, \rho_{s1}(t)] + (N-1)Tr_2[V_{12}, \rho_{s12}(t)] + Tr_{23...N}[V_s(t), \rho_s(t)]$$

$$i\rho_{s12}'(t) = [H_{s1}+H_{s2}+V_{12}, \rho_{s12}(t)]+(N-2)Tr_3[V_{13}+V_{23}, \rho_{s123}(t)]+Tr_{34...N}[V_s(t), \rho_s(t)]$$

These are the basic equations from which the quantum Boltzmann equation is derived making further approximations. We make further approximations:

$$Tr_{23...N}[V_s(t), \rho_s(t)] \approx Tr_{23...N}[V_s(t), \rho_{s1}(t) \otimes ...\rho_{sN}(t)]$$

and

$$V_s(t) \approx \sum_{k=1}^{N} V_{sk}(t)$$

where $V_{sk}(t)$ acts in $\mathcal{H}_k$ and for different $k's$, these are identical copies. Then,

$$Tr_{23...N}[V_s(t), \rho_{s1}(t) \otimes ...\rho_{sN}(t)] \approx$$

$$[V_{s1}(t), \rho_{s1}(t)]$$

The logic in this latter approximation is that the bath field acts in an additive way separtely on each of the system particles and hence can be regarded as an external time varying potential in which each of the system particles moves. Thus, we may write

$$H_{s1}(t) = H_{s1} + V_{s1}(t)$$

and this Hamiltonian acts in $\mathcal{H}_1$, the Hilbert space of the first particle in the system. Further, we approximate $Tr3[V_{13} + V_{23}, \rho_{s123}(t)]$ by

$$Tr3[V_{13} + V_{23}, \rho_{12}(t) \otimes \rho_3(t)]$$

which is obtained on writing

$$\rho_{s123}(t) \approx \rho_{s12}(t) \otimes \rho_3(t) + g_{123}(t)$$

where $g_{123}(t)$ is of the first order of smallness, and neglecting second order of smallness terms. Further we write,

$$\rho_{12}(t) = \rho_1(t) \otimes \rho_2(t) + g_{12}(t)$$

where $g_{12}(t)$ is of the first order of smallness. Then neglecting second order of smallness terms, we have

$$Tr3[V_{13} + V_{23}, \rho_{123}(t)]$$

$$\approx Tr3[V_{13} + V_{23}, \rho_1(t) \otimes \rho_2(t) \otimes \rho_3(t)]$$
$$= [W_1(t) + W_2(t), \rho_1(t) \otimes \rho_2(t)]$$

where

$$W_1(t) = Tr3[V_{13}, \rho_3(t)], W_2(t) = Tr3[V_{23}, \rho_3(t)]$$

Note that $W_1(t)$ and $W_2(t)$ are identical copies of each other acting in the Hilbert spaces $\mathcal{H}_1$ and $\mathcal{H}_2$ respectively. Since we are neglecting second order of smallness terms, we can substitute for $\rho_3(t)$ the expression

$$\rho_3(t) = exp(-it.ad(H_{s3})(\rho_3(0))$$

in the above expressions for $W_1(t), W_2(t)$. If we are interested in calculating $\rho_{s1}(t)$ upto the first order of smallness terms, then we would use the simplified quantum Boltzmann equation

$$i\rho'_{s1}(t) = [H_{s1}(t), \rho_{s1}(t)] + (N-1)[V_{12}, \rho_{s1}(t) \otimes \rho_{s1}(t)]$$

If however, we are interested in calculating $\rho_{s1}(t)$ upto the second order of smallness terms, then we would use

$$i\rho'_{s1}(t) = [H_{s1}(t), \rho_{s1}(t)] + (N-1)[V_{12}, \rho_{s1}(t) \otimes \rho_{s1}(t) + g_{12}(t)]$$

where by first order perturbation theory applied to the above equation satisfied by $\rho_{s12}(t)$, we have

$$ig'_{12}(t) = [H_{s1} + H_{s2}, g_{12}(t)] + [V_{12}, \rho_{s1}(t) \otimes \rho_{s1}(t)]$$

$$+(N-2)[W_1(t)+W_2(t), \rho_{s1}(t)\otimes\rho_{s1}(t)]$$

[34] **Some problems in fluid dynamics related to large deviation theory**

[1] Consider an incompressible fluid described by the following equations for the velocity field:

$$v_x(t,x,y)v_{x,x}(t,x,y)+v_y(t,x,y)v_{x,y}(t,x,y)+v_{x,t}(t,x,y)$$

$$=-p_{,x}(t,x,y)+\eta(v_{x,xx}(t,x,y)+v_{x,yy}(t,x,y))+\sqrt{\epsilon}f_x(t,x,y)$$

$$v_x(t,x,y)v_{y,x}(t,x,y)+v_y(t,x,y)v_{y,y}(t,x,y)+v_{y,t}(t,x,y)$$

$$=-p_{,y}(t,x,y)+\eta(v_{y,xx}(t,x,y)+v_{y,yy}(t,x,y))+\sqrt{\epsilon}f_y(t,x,y),$$

$$v_{x,x}(t,x,y)+v_{y,y}(t,x,y)=0$$

where $(f_x(t,x,y), f_y(t,x,y))$ is a random forcing field. Calculate the rate function for the velocity and pressure fields over a given region of space-time assuming that the forcing field is a Gaussian field with specified mean and covariance.

[2] Let $X_1, X_2, ..$ be iid random variables and define

$$S_n = X_1 + ... + X_n$$

Calculate the limit of the distribution

$$Pr(X_1 \in dx|S_n/n = m)$$

as $n \to \infty$. Make use of Cramer's theorem on large deviations.

hint: Let $I(x)$ denote the rate function of $\{S_n/n : n \geq 1\}$. Then, we have approximately for large n,

$$Pr(X_1 \in dx|S_n/n = m) = P(X_1 \in dx, X_2 + ... + X_n = nm - x)/P(S_n/n = m)$$

$$= P(X_1 \in dx).P((X_2 + ... + X_n)/(n-1) = (nm-x)/(n-1))/P(S_n/n = m)$$

$$= P(X_1 \in dx).exp(-(n-1)I((nm-x)/(n-1))/exp(-nI(m))$$

$$= P(X_1 \in dx).exp(-(n-1)(I(m)+I'(m)x)/exp(-nI(m))$$

$$= P(X_1 \in dx).exp(I'(m)x + I(m))$$

Historical remark: Guo, Papanicolau, Kipnis, Olla and Varadhan introduced and averaging method which enables one to obtain hydrodynamic scaling limits for the interacting diffusion and simple exclusion models when the number of particles tends to infinity and the lattice on which the particles are present converges to the continuum.

Let $\eta_t(x)$ be a simple exclusion process on $\mathbb{Z}$. It satisfies the stochastic differential equations

$$d\eta_t(x) = \sum_{y\neq x}[\eta_t(y)(1-\eta_t(x))dN_t(y,x) - \eta_t(x)(1-\eta_t(y))dN_t(x,y)]$$

where $\{N_t(x,y) : t \geq 0\}_{x\neq y}$ are independent Poisson processes with means of

$$\mathbb{E}(N_t(x,y)) = p(x,y)t$$

Thus, the generator of the η_t process is given by

$$\mathbb{E}[\phi(\eta_{t+dt})|\eta_t] = \phi(\eta_t) + dt.L\phi(\eta_t)$$

where for $\eta : \mathbb{Z} \to \{0,1\}$,

$$L\phi(\eta) = \sum_{x\neq y} p(x,y)\eta(x)(1-\eta(y))(\phi(\eta^{(x,y)}) - \phi(\eta))$$

Let us assume that

$$p(x,y) = p(y-x)$$

Then, we get

$$L\phi(\eta) = \sum_{x,z} p(z)\eta(x)(1-\eta(x+z))(\phi(\eta^{(x,x+z)}) - \phi(\eta))$$

Define the empirical distribution of the particles at time t by

$$\mu_{N,t} = N^{-1}\sum_x \eta_t(x)\delta_{x/N}$$

where the sum is over $x \in \mathbb{Z}_N = \{0,1,...,N-1\}$. Note that μ_N is a random measure on $\mathbb{Z}_N$. By identifying $x \in \mathbb{Z}_N$ with $x/N = \theta \in [0,1]$, we can regard $\mu_{N,t}$ as a random measure on $[0,1]$. Then, for any function $J : [0,1] \to \mathbb{R}$, we have that

$$\int_0^1 J(\theta)d\mu_{N,t}(\theta) = (1/N)\sum_x J(x/N)\eta_t(x) = \xi_{J,N}(t)$$

say. Then,

$$d\xi_{J,N}(t) = (1/N)\sum_x J(x/N)d\eta_t(x) =$$

$$(1/N)\sum_{x\neq y} J(x/N)(\eta_t(y)(1-\eta_t(x))dN_t(y,x) - \eta_t(x)(1-\eta_t(y))dN_t(x,y))$$

$$= (1/N)\sum_{x\neq y} J(x/N)(\eta_t(y)(1-\eta_t(x))p(x-y) - \eta_t(x)(1-\eta_t(y))p(y-x))dt$$

$$+dM_{J,N}(t)$$

$$= dt(1/N)\sum_{x=y}(J(x/N) - J(y/N))p(x-y)\eta_t(y)(1-\eta_t(x)) + dM_{J,N}(t)$$

where $M_{J,N}(t)$ is a Martingale defined by

$$dM_{J,N}(t) = (1/N)\sum_{x\neq y} J(x/N)(\eta_t(y)(1-\eta_t(x))dM_t(y,x) - \eta_t(x)(1-\eta_t(y))dM_t(x,y))$$

$$= (1/N)\sum_{x\neq y}(J(x/N) - J(y/N))\eta_t(y)(1-\eta_t(x))dM_t(y,x)$$

where

$$M_t(y,x) = N_t(y,x) - p(x-y)t, x \neq y$$

are independent Martingales. We note that

$$\mathbb{E}[(dM_{J,N}(t))^2] = (dt/N^2)\sum_{x\neq y}(J(x/N) - J(y/N))^2\mathbb{E}((\eta_t(y)(1-\eta_t(x)))^2 p(x-y)^2 dt$$

which is easily seen to tend to zero as $N \to \infty$. In fact, it is bounded above by

$$(dt/N^2)(sup_{\theta\in[0,1]}J'(\theta)^2)\sum_{x,y\in\mathbb{Z}_N}(x-y)^2 p(x-y)/N^2$$

$$= (dt/N^3)(sup_{\theta\in[0,1]}J'(\theta)^2)\sum_z z^2 p(z)$$

which will converge to zero in the special case when $p(z)$ has finite range or more generally even when $\sum_z p(z)$ is finite which is then always the case since $p(z)$ equals λ times the probability that a particle will jump from x to $x+z$ where λ is the rate of the Poisson clock. Note that

$$\sum_{z\in\mathbb{Z}_N} z^2 p(z) \leq N^2\sum_z p(z) = \lambda N^2$$

Thus, with neglect of the martingale terms, we find that

$$d\xi_{J,N}(t) = dt(1/N)\sum_{x \neq y}(J(x/N) - J(y/N))p(x-y)\eta_t(y)(1-\eta_t(x))$$

$$= (dt/N)\sum_{y,z}(J'(y/N)z/N + J''(y/N)z^2/2N^2)p(z)\eta_t(y)(1-\eta_t(y+z))$$

with neglect of terms which converge to zero as $N \to \infty$. For symmetric dynamics ($p(z) = p(-z)$), $\sum_z zp(z) = 0$ and if we assume that $\sum_z z^2 p(z) = DN^2$, then we can prove by the method of averaging due to Guo, Papanicolau and Varadhan that

$$\sum_y F(y)\eta(y)(1-\eta(y+z))$$

can be replaced by

$$\sum_y F(y)(2N\epsilon+1)^{-1}\sum_{x:|x-y|\leq N\epsilon}\eta(x)(1-\eta(x+z))$$

in the limit when $N\to\infty$ and then $\epsilon\to 0$. This is called the principle of local averaging. From this fact, one is able to prove that the limiting probability distribution of the particles is concentrated on a set where the local density is non-random and satisfies Burger's partial differential equation.

Remarks: Let

$$\bar{\eta}_{k,x}=(2k+1)^{-1}\sum_{y:|y-x|\leq k}\eta(y)$$

Then, we have that if $g(\theta)$ is smooth function on $[0,1]$,

$$N^{-1}\sum_{x\in\mathbb{Z}_N}g(x/N)\eta(x)$$

can be replaced by

$$N^{-1}\sum_x g(x/N)\bar{\eta}_{k,x}$$

for fixed large k as $N\to\infty$. This is because

$$|N^{-1}\sum_x g(x/N)\eta(x)-N^{-1}\sum_x g(x/N)\bar{\eta}_{k,x}|$$

$$=N^{-1}|\sum_x g(x/N)\eta(x)-\sum_y \eta(y)(2k+1)^{-1}\sum_{x:|x-y|\leq k}g(x/N)|$$

$$\leq N^{-1}\sum_y|g(y/N)-(2k+1)^{-1}\sum_{x:|x-y|\leq k}g(x/N)|\eta(y)$$

Now for any $\epsilon>0$, for fixed k, we have that for all $N>N_0(\epsilon,k)$

$$sup_y|g(y/N)-(2k+1)^{-1}\sum_{x:|x-y|\leq k}g(x/N)|<\epsilon$$

since a continuous function on a compact interval is also uniformly continuous. Thus, we get

$$lim_{N\to\infty}N^{-1}\sum_x g(x/N)\eta(x)-N^{-1}\sum_x g(x/N)\bar{\eta}_{k,x}=0$$

More generally, we can replace

$$(1/N)\sum_x g(x/N)F(\tau_x\eta)$$

in view of the regularity of g on $[0,1]$ by

$$(1/N)\sum_x g(x/N)(2k+1)^{-1}\sum_{y:|y-x|\leq k} F(\tau_y\eta)$$

and we can then replace $(2k+1)^{-1}\sum_{y:|y-x|\leq k} F(\tau_y\eta)$ by $\hat{F}((2k+1)^{-1}\sum_{y:|y-x|\leq k}\tau_y\eta)$ if we assume that the $\eta(x)'s$ are independent Bernoulli random variables with corresponding means $\rho(x/N)$ and we define

$$\hat{F}(\rho) = \mathbb{E}_\rho F(\eta)$$

where if $\eta = (\eta(x) : x \in \mathbb{Z}_N)$, then $\mathbb{E}_\rho F(\eta)$ is the mean of $F(\eta(x) : x \in \mathbb{Z}_N)$ with $\eta(x)'s$ being independent Bernoulli with means $\rho(x)'s$. This is an elementary consequence of the law of large numbers. It is valid provided that we take k sufficiently large. In proving hydrodynamic scaling limits, the kind of $F(\eta)$ that we encounter is typically $F(\eta) = \eta(0)(1-\eta(1))$ which results in $F(\tau_x\eta) = \eta(x)(1-\eta(x+1))$ and then

$$(1/N)\sum_x g(x/N)\eta(x)(1-\eta(x+1))$$

can be replaced by

$$(1/N)\sum_x g(x/N)\rho(x/N)(1-\rho((x+1)/N))$$

from which the hydrodynamical scaling limit for nearest neighbour interactions can be derived. More generally, if the jump probabilities $p(z)$ have finite range, then

$$(1/N)\sum_{x,z} g(x/N)p(z)\eta(x)(1-\eta(x+z)) - - - - - (a)$$

can be represented by

$$(1/N)\sum_x g(x/N)F(\tau_x\eta)$$

where

$$F(\eta) = \sum_z p(z)\eta(0)(1-\eta(z))$$

and then we get that (a) can be replaced by

$$(1/N)\sum_x g(x/N)p(z)\rho(x/N)(1-\rho((x+z)/N)$$

which immediately leads to the hydrodynamical scaling limit which is clearly linear in the symmetric case, ie, when $p(z) = p(-z)$.

[35] Some historical remarks: The RLS lattice algorithm involving computation of the prediction filter coefficients recursively both in order and time was

first accomplished by the Greek engineers "Carayannis,Manolakis and Kalouptsidis" in a pioneering paper published in the the IEEE transactions on Signal processing sometime during the eighties. It was polished further and stability analysis carried out by the team of Thomas Kailath, an Indo-American engineer. Thomas Kailath and his team members also formulated the RLS lattice algorithm for multidimensional time series models. The final algorithm for the RLS lattice algorithm for multivariate time series with application to cyclostationary process prediction was carried out by Mrityunjoy Chakraborty in his Ph.D thesis. A.Paulraj invented the ESPRIT algorithm for estimating the frequencies in a noisy harmonic process based on rotationally invariant techniques. This work appeared in the IEEE transactions on signal processing sometime in the mid eighties along with Richard Roy and Thomas Kailath. The idea was a computationally efficient high resolution eigensubspace based algorithm for estimating the sinusoidal frequencies or equivalently the directions of arrival of multiple plane wave signals. It revolutionized the entire defence and astronomy industry. Later on higher dimensional versions of the MUSIC and the ESPRIT algorithm were invented by the author along in his PhD thesis along with his supervisors. These contributions are present in the papers

[1] Harish Parthasarathy, S.Prasad and S.D.Joshi, "A MUSIC like method for bispectrum estimation", Signal Processing, Elsevier, North Holland, 1994.

[2] Harish Parthasarathy, S.Prasad and S.D.Joshi, "An ESPRIT like algorithm for quadaratic phase coupling estimation", IEEE transactions on Signal Processing, 1995.

The scientific contributions of Subramaniyam Chandrasekhar:

Chandraskehar as college student, in the early 1940's discovered a remarkable consequence of combining relativity, gravitation and quantum mechanics:by applying ideas from these newly developed fields to study the equilibrium of a star that has exhausted it fuel and is contracting under the influence of gravitation with a counter repulsive force caused by the pressure exerted by the degenerate electron gas within the star owing to the Paul exclusion principle. In this way, Chandraskehar arrived at a fundmental limiting radius of a star that has exhausted all its fuel. Such a star is called a white dwarf and Chandrasekhar's calculations showed that this limiting radius is around 1.5 times the mass of the sun. The idea behind Chandraskehar's calculation can be understood as follows. All the electrons in the star after its fuel has been exhausted have energies below the Fermi level and hence if $n(p)$ denotes the number of electrons per unit volume in phase space and m_e the rest mass of an electron, then the pressure of this degenerate electron gas is from the basic formula for the energy-momentum tensor

$$T^{\mu\nu} = \sum_a (P_a^\mu P_a^\nu / E_a)\delta^3(x - x_a)$$

given by (note that $\int T^{\mu\nu} d^4x$ is Lorentz invariant)

$$P = \int_{p^2/2m_e < E_F} n(p)(p^2/E(p))d^3p$$

Noting that $n(p) = 1/h$, we get

$$P = \int_0^{\sqrt{2m_e E_F}} p^2.4\pi p^2 dp/\sqrt{m_e^2c^4 + p^2c^2} = P(E_F)$$

say. On the other hand, the volume density of this degenerate electron gas is given by

$$\rho = \int_{p^2/2m_e<E_F} (E(p)/c^2)n(p)d^3p = c^{-2}\int_0^{\sqrt{2m_e E_F}} \sqrt{m_e^2c^4 + p^2c^2}.4\pi p^2 dp = \rho(E_F)$$

Chandraskehar then eliminated E_F between these to equations and obtained the equation of state $P = P(\rho)$ of this degenerate electron gas. He showed that this equation could well be approximated by a "polytrope":

$$P = C\rho^\gamma, \gamma \approx 4/3$$

where the constant C is a function of the fundamental constants c, h, m_e.

On the influence of group theory and group representation theory developed by Harish-Chandra on problems in statistical image processing.

On the influence of the book on Lie groups, Lie algebra and their representations by V.S.Varadarajan on robotics.

Varadarajan in his book develops all the analytic and algebraic tool required for constructing all the irreducible representations of a complex semisimple Lie algebra/Lie group using the method of the quotient of universal enveloping algebra by a maximal ideal as first developed by Harish-Chandra in his celebrated paper "On some applications of the universal enveloping algebra of a semisimple Lie algebra." Prior to this topic, Varadarajan discusses all the other important analytic tools in Lie group theory like the differential of the exponential map, the Baker-Campbell-Hausdorff formula for the product of the exponential of two Lie algebra elements (for example in the matrix/linear Lie group case, the product of the exponential of two matrices). These ideas play a fundamental role in robotics where for example, the robot comprises of connected three dimensional links, each of which can undergo translation and rotation. Thus, the configuration space of the entire robot is the direct product of n elements of a subgroup of the full three dimensional Euclidean motion group. Varadarajan then gives a rigorous proof of Weyl's celebrated character formula for all compact semisimple Lie groups in terms of the dominant integral weight of the representation first introduced by Elie Cartan. This idea is fundamental and can be applied to problems of image processing as follows. Suppose that we are given a manifold $\mathcal{M}$ on which the compact group G acts. An element of G transforms an image field $f : \mathcal{M} \to \mathbb{C}$ into another image field $f_1(x) = f(g^{-1}x), x \in \mathcal{M}$. Now we wish to associate an "invariant functional" of the image field that will yield the same number for f as well as f_1 no matter what the element g of G is. This is possible by taking an irreducible character $\chi(g)$ of G and constructing the functional

$$I(f) = \int_{G\times G} \chi(h^{-1}k)f(kx_0)f(hx_0)dhdk$$

For example then

$$I(f_1) = \int_{G\times G} \chi(h^{-1}k)f(g^{-1}kx_0)f(g^{-1}hx_0)dhdk$$

$$= \int_{G\times G} \chi(h^{-1}k)f(kx_0)f(hx_0)dhdk = I(f)$$

So for each irreducible representation of G we have an associated invariant functional and Weyl's character formula combined with Weyl's integration formula (which reduces an integral of a function on the group to an integral on a Cartan subgroup of the orbital integral of the function.

On the influence of the book "Perturbation theory for linear operators" by Tosio Kato on developments in robotics and quantum mechanics. Kato single-handedly created the notion of relative boundedness and relative compactness of an operator w.r.t. another in Banach and Hilbert spaces and applied it to a systematic study of unbounded operators in Hilbert space and also to obtain results like when the perturbation of an unbounded closed opearator by another one remains closed. This notion is also known as stability of closedness. He also applied the notion of relative boundedness to answer the question of when the perturbation of an unbounded self-adjoint operator by a closed symmetric operator remains bounded. His book is a masterpiece as the only analytic tool that he uses in answering most of such questions about unbounded operators is based on the resolvent of the operator and its integral around a closed curve in the complex plane after multiplying it by a function of the complex variable. Kato's book is applicable to a variety of problems in quantum mechanics such as obtaining bounds on the charge of the electron that would guarantee that the Hamiltonian of the two electron atom is self-adjoint.

On the influence of the papers of Albert Einstein on developments in modern cosmology, evolution of inhomogeneities in the expanding universe and in the unification of gravity with quantum mechanics.

On the influence of the books of Steven Weinberg on modern quantum field theory especially in the problem of electroweak unification, ie, unification of the weak nuclear forces with electromagnetism.

The volumes start with the canonical quantization of the Klein-Gordon field based on transforming the Lagrangian density of the field to a Hamiltonian density and introducing canonical commutation relations between the position and momentum fields (equal time commutation relations) and then by means of a spatial Fourier decomposition of this field in terms of creation and annihilation operator fields in momentum space, the author derives the canonical commutation relations for these operators. Prior to this, the author introduces the notion of a wave function in second quantized Hilbert corresponding to a definite four momentum and spin/helicity space and how a unitary representation of the Lorentz group acts on such a wave function. The action of such a representation turns out from natural principles to have two components, one a

Lorentz transformation of the four momentum of the wave function and two a little group element acting on the helicity variables. The author then presents the basic elements of quantum scattering theory. This presentation is based on assuming that the free particle state evolving under the free projectile Hamiltonian first interacts with the scattering centre causing it to get scattered to an in-scattered state evolving under the free particle Hamiltonian plus an interaction potential in such a way that in the remote past, both the free particle and the interacting particle wave functions coincide. This gives rise to the notion of a wave operator. Likewise the particle gets scattered from the in scattered state to an out scattered state and then gets scattered to a free particle state as time goes to infinity in such a way that the out scattered state evolving under the interaction Hamiltonian coincides as time goes to infinity to a free particle out state evolving under the free particle Hamiltonian. Thus we get two kinds of wave operators and one constructs the scattering matrix using these two operators. Weinberg with a remarkable physical insight skips over rigorous mathematical details involving the various notions of operator convergence and directly gives us the relationship between the inparticle free state and the corresponding inparticle scattered state in terms of the interaction potential and the spectrum of the free particle Hamiltonian. Likewise, he tells us the corresponding relationship between the out free particle and out scattered states. These relationships can be made mathematically rigorous using the spectral theorem and spectral measures for unbounded operators in Hilbert space and are known as the Lippmann-Schwinger equations. See for example the classic "Perturbation theory for linear operators" by Tosio Kato.

Salam and Strathdhee introduced the notion of a super-vector field as a first order differential operator in the Bosonic and Fermionic coordinates. They used such supervector field to define infinitesimal transformations of superfields under a supersymmetry transformation. Later on, researchers in quantum field theory showed how to write down various kinds of supersymmetric actions, ie, action functionals that remain invariant under supersymmetry transformations apart from remaining invariant under also Lorentz and gauge transformations. These later developments are marvellously described in the third volume on "Supersymmetry" by Steven Weinberg.

[36] **Remarks about supergravity**

This section of the book has been adapted from celebrated volumes of Michael Green, John Schwarz and Edward Witten on Superstring Theory. Some part has also been taken from Weinberg, Vol.III.

The most general form of a supergravity action is that it contains apart from the Bosonic part involving the Einstein Hilbert action functional of the tetrad field, Fermionic parts involving the gravitino. The gravitino is the superpartner of the graviton. The tetrad or Vierbein field describes the gravition while the gravitino field has both a spinor index and a vector index and is therefore a particle of spin 3/2. The supergravity field also contains a Yang-Mills gauge field component. This gauge field is a Bosonic field and therefore the action should also contain the super-partner of this Yang Mills field namely a Fermionic spin 1/2 field which is the usual matter field action of the Dirac field generalized to the Yang-Mills gauge group. The whole action must be invariant under local supersymmetry transformations apart from being invariant under local Lorentz transformations, diffeomorphisms of space-time and gauge transformations and it remains therefore to define such local supersymmetry transformations in such a way that the entire set of all the above mentioned four kinds of transformations closes on itself, ie, forms a Lie algebra of transformations. The starting point for constructing an appropriate action for supergravity is the equation of supercurrent conservation.

Following Green, Schwarz and Witten, we write a simplified form of the supergravity Lagrangian without involving the Yang Mills matter and gauge terms as

$$L = e[c_1 R + \bar{\chi}_\mu \Gamma^{\mu\nu\rho} D_\nu \chi_\rho]$$

where

$$D_\nu = \partial_\nu + \omega_\nu^{mn} \Gamma_{mn}$$

and

$$R = R_{\mu\nu}^{mn} e_m^\mu e_n^\nu,$$

$$R_{\mu\nu}^{mn} = \omega_{\nu,\mu}^{mn} - \omega_{\mu,\nu}^{mn} + [\omega_\mu, \omega_\nu]^{mn}$$

Note that χ_μ or equivalently χ_a is a Majorana Fermion which means that

$$\bar{\chi}_\mu = \chi_\mu^* \Gamma^0 = \chi^{\mu T} \Gamma^5 \epsilon$$

so that

$$\chi^{\mu *} = \chi^{\mu T} \gamma^5 \epsilon \Gamma^0$$

is the adjoint (ie, conjugate transpose) of χ_μ. Under the local supersymmetry transformation defined by the Majorana Fermionic parameter field $\epsilon(x)$, we have

$$\delta \chi_\mu = D_\mu \epsilon(x), \delta e_\mu^a = \Gamma^a \chi_\mu(x)$$

and the transformation law of ω_μ^{mn} is determined by the field equation that it satisfies, ie,

$$-\delta\omega_\nu^{mn}(e_m^\mu e_n^\nu)_{,\mu} + \delta\omega_\mu^{mn}(e_m^\mu e_n^\nu)_{,\nu} + ([\delta\omega_\mu, \omega_\nu]^{mn} + [\omega^\mu, \delta\omega_\nu]^{mn}) e_m^\mu e_n^\nu$$

$$+ \bar{\chi}_\mu \Gamma^{\mu\nu\rho} \Gamma_{mn} \chi_\rho \delta\omega_\nu^{mn} = 0$$

Remark:

$$[\omega_\mu, \omega_\nu]^{mn} \Gamma_{mn} = \omega_\mu^{pq} . \omega_\nu^{rs} [\Gamma_{pq}, \Gamma_{rs}]$$

where

$$[\Gamma_{pq}, \Gamma_{rs}] = \Gamma_{ps}\eta_{qr} + \Gamma_{qr}\eta_{ps} - \Gamma_{pr}\eta_{qs} - \Gamma_{qs}\eta_{pr}$$

from which we deduce that

$$[\omega_\mu, \omega_\nu]^{mn} = [\omega_\mu^{mq} \omega_\nu^{rn} \eta_{qr}$$

$$+ \omega_\mu^{pm} \omega_\nu^{ns} \eta_{ps} - \omega_\mu^{mq} \omega_\nu^{ns} \eta_{qs}$$

$$-\omega_\mu^{pm}\omega_\nu^{rn}\eta_{pr}]$$

$$= -4\omega_\mu^{mp}\omega_\nu^{nq}\eta_{pq}$$

This equation implies that for all variations $\delta\omega_\mu^{mn}$, the following equation is satisfied:

$$-\delta\omega_\mu^{mn}(e_m^\nu e_n^\mu)_{,\nu}+\delta\omega_\mu^{mn}(e_m^\mu e_n^\nu)_{,\nu}-2\delta\omega_\mu^{mp}\omega_\nu^{nq}\eta_{pq}e_m^\mu e_n^\nu+\bar{\chi}_\mu\Gamma^{\nu\mu\rho}\Gamma_{mn}\chi_\rho\delta\omega_\mu^{mn}=0$$

Thus, ω_μ^{mn} satisfies the algebraic field equation:

$$-(e_m^\nu e_n^\mu)_{,\nu}+(e_m^\mu e_n^\nu)_{,\nu}-2\omega_\nu^{pq}\eta_{nq}e_m^\mu e_p^\nu+\bar{\chi}_\nu\Gamma^{\nu\mu\rho}\Gamma_{mn}\chi_\rho=0$$

It is an easy matter to solve this algebraic field equation for ω_μ^{mn}.

Now the quadratic part in ω_μ^{mn} when the action is varied under the above supersymmetry transformation is

$$-4c_1\omega_\mu^{mp}\omega_\nu^{nq}\eta_{pq}+\omega_\mu^{mn}\epsilon_0(x)^T\Gamma_{mn}^T\Gamma_5\epsilon\Gamma^{\mu\nu\rho}\omega_\nu^{rs}\chi_\rho$$

$$+\chi_\mu^T\Gamma_5\epsilon.\Gamma^{\mu\nu\rho}\Gamma_{rs}\Gamma_{mn}\omega_\nu^{rs}\omega_\rho^{mn}\epsilon_0(x)$$

[37] One of the main achievements in the work of C.R.Rao was the proof of the lower bound on the error covariance matrix of a statistical estimator of a vector valued parameter based on vector valued observations using techniques of matrix theory. C.R.Rao in his work has also considered the case when the Fisher information matrix is singular and in this case, he has been able to use the methods of generalized inverses to obtain new formulas for the lower bound. The lower bound on the variance of an estimator should be compared to the Heisenberg uncertainty principle in quantum mechanics for two non-commuting observables. In fact, it can be shown that the Heisenberg uncertainty inequality for position and momentum can be derived using the CRLB. The CRLB roughly tells us that no matter how much we may try, we can never achieve complete accuracy in our estimation process, ie, there is inherently some amount of uncertainty about the system that generates a random observation.

Remarks about efficiency of a statistical estimator of a parameter using CRLB

Let $X \in \mathbb{R}^n$ be the vector valued measurement and $\theta \in \mathbb{R}^p$ be the parameter to be estimated. The pdf of the measurement is $p(X|\theta)$. Let $\hat{\theta}(X)$ be any estimator of θ based on the measurement X. The bias of this estimator is

$$B(\theta) = \mathbb{E}(\hat{\theta}(X)) - \theta$$

which can be alternately expressed as

$$\int (\hat{\theta}(X) - \theta)p(X|\theta)dX = B(\theta)$$

or in component form,

$$\int (\hat{\theta}_a(X) - \theta_a)p(X|\theta)dX = B_a(\theta), a = 1, 2, ..., p$$

Differentiating both sides of this expression w.r.t θ_b and using $\int p(X|\theta)dX = 1$ gives

$$-\delta_{ab} + \int (\hat{\theta}_a(X) - \theta_a)(\partial p(X|\theta)/\partial\theta_b)dX = \partial B_a(\theta)/\partial\theta_n b$$

This equation is multiplied by $u_a v_b$ and summed over $a, b = 1, 2, ..., p$ to get

$$-u^T v + \mathbb{E}[u^T(\hat{\theta}(X) - \theta).\nabla_\theta ln(p(X|\theta)^T v] = u^T B'(\theta)v$$

or equivalently,

$$\mathbb{E}[u^T(\hat{\theta}(X) - \theta).(\nabla_\theta ln(p(X|\theta)))^T v] = u^T(I + B'(\theta))v$$

from which we deduce using the Cauchy-Schwarz inequality,

$$(u^T(I + B'(\theta))v)^2 \leq \mathbb{E}[(u^T(\hat{\theta}(X) - \theta))^2].\mathbb{E}[(v^T\nabla_\theta lnp(X|\theta)^2]$$

$$= (u^T Ru)(v^T Jv)$$

where

$$R = \mathbb{E}[(\hat{\theta}(X) - \theta).(\hat{\theta}(X) - \theta)^T]$$

and

$$J = \mathbb{E}(\nabla_\theta lnp(X|\theta).(\nabla_\theta lnp(X|\theta))^T)$$

$$= ((\mathbb{E}(\frac{\partial lnp(X|\theta)}{\partial\theta_i}.\frac{\partial lnp(X|\theta)}{\partial\theta_j})))$$

Note that if $\hat{\theta}(X)$ is an unbiased estimator of θ, then $B(\theta) = 0$ and

$$R = Cov(\hat{\theta}(X)) = C$$

and the above inequality reduces to

$$(u^T v)^2 \leq (u^T Cu)(v^T Jv)$$

In the general case, consider

$$(u^T Gv)^2 \leq (u^T Ru)(v^T Jv), G = I + B'(\theta), B'(\theta) = \nabla_\theta B(\theta)$$

Substitute into this

$$v = Au$$

to get

$$(u^T GAu)^2 \leq (u^T Ru).(u^T A^T JAu)$$

This inequality holds for any $p \times p$ real matrix A. Let in particular,

$$A = J^{-1}G^T$$

to get

$$(u^T G J^{-1} G^T u)^2 \leq (u^T R u).(u^T G J^{-1} G^T u)$$

or equivalently,

$$u^T G J^{-1} G^T u \leq u^T R u$$

or equivalently,

$$u^T (R - G J^{-1} G^T) u \geq 0 for all u \in \mathbb{R}^p$$

In other words,

$$R \geq G J^{-1} G^T$$

in the sense that the difference between the LHS and the RHS is a positive semidefinite matrix. In particular, if we allow u to run over an orthonormal basis for $\mathbb{R}^p$ and add all the inequalities so obtained, we get

$$Tr(R) \geq Tr(G J^{-1} G^T)$$

which is the same as saying that

$$\mathbb{E}[\| \hat{\theta}(X) - \theta \|^2 \geq Tr(G J^{-1} G^T)$$

The lhs in this inequality is the mean square estimation error. In the special case, when the estimator is unbiased, this inequality reads

$$Var(\hat{\theta}(X)) \geq Tr(J^{-1})$$

Remark: There is an alternate expression for the Fisher information matrix:

$$J_{ij} = \mathbb{E}[(\partial lnp/\partial\theta_i).(\partial lnp/\partial\theta_j)]$$

$$= \int p(lnp)_{,i}.(lnp)_{,j} dX = \int p_{,i}(lnp)_{,j} dX$$

$$= \partial_i \int p(lnp)_{,j} dX - \int p(lnp)_{,ij} dX$$

$$= \partial_i \int p_{,j} dX - \mathbb{E}(lnp)_{,ij}$$

$$= \partial_i \partial_j \int p dX - \mathbb{E}(lnp)_{,ij} = -\mathbb{E}(lnp)_{,ij}$$

where ∂_i, $Z_{,i}$ etc, denote differentiation w.r.t θ_i and not w.r.t X_i. Thus,

$$J_{ij} = -\mathbb{E}[\frac{\partial lnp(X|\theta)}{\partial\theta_i \partial\theta_j}]$$

For Product Safety Concerns and Information please contact our EU representative GPSR@taylorandfrancis.com
Taylor & Francis Verlag GmbH, Kaufingerstraße 24, 80331 München, Germany

www.ingramcontent.com/pod-product-compliance
Lightning Source LLC
Chambersburg PA
CBHW060300220326
41598CB00027B/4185
9781032405162